T0222841

THE DARK MATTER PROBLEM

A Historical Perspective

Most astronomers and physicists now believe that the matter content of the Universe is dominated by dark matter: hypothetical particles which interact with normal matter primarily through the force of gravity. Though invisible to current direct detection methods, dark matter can explain a variety of astronomical observations. This book describes how this theory has developed over the past 75 years, and why it is now a central feature of extragalactic astronomy and cosmology.

Current attempts to directly detect dark matter locally are discussed, together with the implications for particle physics. The author comments on the sociology of these developments, demonstrating how and why scientists work and interact. Modified Newtonian Dynamics (MOND), the leading alternative to this theory, is also presented.

This fascinating overview will interest cosmologists, astronomers, and particle physicists. Mathematics is kept to a minimum, so the book can be understood by non-specialists.

ROBERT H. SANDERS is Professor Emeritus at the Kapteyn Astronomical Institute, Groningen, the Netherlands. He has worked in the field of dark matter for many years.

THE DARK MATTER PROBLEM

A Historical Perspective

ROBERT H. SANDERS

Kapteyn Astronomical Institute, Groningen

CAMBRIDGE
UNIVERSITY PRESS

CAMBRIDGE
UNIVERSITY PRESS

The Edinburgh Building, Cambridge CB2 8RU, UK

Published in the United States of America by Cambridge University Press, New York

Cambridge University Press is part of the University of Cambridge.

It furthers the University's mission by disseminating knowledge in the pursuit of education, learning and research at the highest international levels of excellence.

www.cambridge.org
Information on this title: www.cambridge.org/9781107677180

First published 2010
First paperback edition 2013

A catalogue record for this publication is available from the British Library

Library of Congress Cataloguing in Publication data
Sanders, Robert H.
The dark matter problem : a historical perspective / Robert H. Sanders.
p.   cm.
Includes bibliographical references and index.
ISBN 978-0-521-11301-4
1. Dark matter (Astronomy)–History.   I. Title.
QB791.3.S25   2010
523.1′126–dc22
2010004532

ISBN 978-0-521-11301-4 Hardback
ISBN 978-1-107-67718-0 Paperback

Contents

Acknowledgements

The task of describing the historical development of the dark matter problem would have been much more difficult without the assistance of a number of colleagues and old friends. In the first place I thank Mort Roberts for numerous conversations and emails on the events surrounding the early observations of spiral-galaxy rotation curves. The insights of such a major participant in these developments have been a treasure house, although I hasten to add that the conclusions drawn here from these insights (particularly concerning matters of priority) have been entirely my own.

Throughout the years I have benefited from uncountable conversations with my Groningen colleagues, Renzo Sancisi and Tjeerd van Albada, on the interpretation of the observations of rotation curves. I deeply value this contact, and their numerous useful suggestions contributed both to the content and presentation of this book. I have also benefited enormously from collaborations with generations of Groningen students: Kor Begeman, Adrick Broeils, Marc Verheijen, Edo Noordermeer, and Rob Swaters. We have spent many hours together looking at and discussing rotation curves of galaxies ranging from faint dwarfs (barely a smudge on the photographic plate) to those of giant luminous spiral systems. Altogether, in their objective and critical way, they have formed my view of the phenomenology.

There is nothing like a discussion with a critical but well-meaning colleague and friend to clarify ideas and sharpen arguments. In this regard I am very grateful to Jacqueline van Gorkom for chats about the sociology of rotation-curve observers and the philosophy of science. I also thank Rien van de Weygaert and Saleem Zaroubi for attempting to fill in the many gaps in my knowledge of cosmology and structure formation, particularly with respect to the historical developments in this field.

Moti Milgrom and Stacy McGaugh have commented on several of the chapters here, and, as always, their remarks are perceptive and helpful. For me it

vii

has been a special privilege to have known and worked with Moti, Stacy and Jacob Bekenstein during the past quarter-century; they are creative and bold scientists.

I would like to thank Vince Higgs and his colleagues at Cambridge University Press for their constant help and many useful suggestions.

Further I am very grateful to Stefano Casertano, Jacob Bekenstein, and especially the French translator, Benoit Clenet, for pointing out embarrassingly many omissions, ambiguities and errors in the first printing of this book.

Finally, I thank my wife, Christine, for her patience and for coming to terms with our different dreams of retirement.

1

Introduction

My purpose here is to discuss the past and present of the dark matter hypothesis: how it has developed that most astronomers and physicists now believe that the matter content of the Universe is dominated by an unseen, non-luminous substance that interacts with ordinary matter, protons, neutrons and electrons, primarily through the force of gravity. This description is personal and based largely upon my perspective as an interpretive astronomer. It is also necessarily biased. Throughout most of my career, for the past 40 years now, I have been involved – at times peripherally, often directly – in research on the discrepancy between the detectable mass of astronomical systems and the inferred Newtonian dynamical mass. Since my graduate student days, I have worked at institutes where consideration of this problem, both theoretical and observational, has been a dominant theme. My views on these developments are certainly colored by my experience at these particular institutes and, no doubt, by my own prejudices. But I do hope that the account that I will give here is reasonably honest and fair.

Forty years ago, I was a graduate student at Princeton University. In the Peyton Hall basement, every Wednesday, there was a lively lunch meeting attended by staff members and students. Theses projects would be described, new ideas would be tossed out and batted around, and often politics (in that lively rebellious period) would be discussed in a highly dialectical manner. One Wednesday – it must have been in 1969 – one of our young assistant professors, Jerry Ostriker, appeared at lunch with a radical new idea. Jerry was an expert on the stability of rotating fluid spheroids (and many other subjects as well). He had been following with interest the computer simulations of disk galaxies which, at that point, were becoming extremely sophisticated, involving large numbers of particles all interacting gravitationally. He had noticed that in these simulations disks of particles which were initially supported against gravity by rotation – let's say, centrifugal force – did not seem to remain that way. The round disks developed elongated shapes and heated up – that is, they became more like hot pressure-supported systems rather than rotating systems.

This corresponded perfectly to what Jerry knew about rotating fluid spheroids: it is impossible to construct such an object supported entirely by rotation; Newtonian dynamical systems supported by rotation are unstable. But our galaxy, the Milky Way Galaxy, appears to be held up almost entirely by rotation; the stars near the Sun are moving on nearly circular orbits about the center of the Galaxy. How is it that the Galaxy can remain rotationally supported and yet stable? Jerry's brilliant leap was to suggest that the Galaxy, in fact, is not rotationally supported – that the rotationally supported disk is only one component of the Galaxy. There is another major component, a spheroidal component, at least equal in mass to the disk, and this system is primarily pressure supported. Because no such massive spheroidal component is seen, it must be dark – a dark halo.

On that Wednesday, this suggestion appeared radical; I recall that it caused a great stir and considerable argument, especially from some of the more senior staff members such as Martin Schwarzschild. He raised a number of questions, most of which concerned the composition of the dark halo (Schwarzschild was an astronomer after all). What is the dark halo made out of? Low-luminosity stars possibly – red dwarfs – remnants of dead stars – white dwarfs. How might it be detected by means other than its gravitational influence? An infrared glow around galaxies, perhaps; high-velocity, low-luminosity stars, maybe. No one could have supposed at that point that the halo might consist of weakly interacting, subatomic particles. This would have been far too radical. Not one of us would have dared to suggest, even if they had thought of it, that Newton's laws might need revision on the scale of galaxies and larger; that would have seemed insane.

In 1973, Ostriker, joined by his Princeton colleague, Jim Peebles, published this proposal which by that point had been bolstered by their own N-body calculations; the idea provoked even more controversy in the larger community than it had on that Wednesday afternoon in Princeton (Chapter 3). Although this was a radically new idea with an entirely theoretical basis, there had been considerable earlier evidence that astronomical systems contain large quantities of unseen matter. In 1933 the Swiss astronomer Fritz Zwicky had made the first systematic kinematic study of a cluster of galaxies and pointed out that in order to gravitationally bind the cluster the actual mass had to be several hundred times larger than the observed mass in stars (Chapter 2). Earlier, in 1932, the Dutch astronomer Jan Oort, by looking at the motion of the stars above the galactic plane, concluded that there must be about 50% more mass in the Galaxy disk than is evidenced by luminous stars. But Oort's dark matter was distributed in the plane of the Galaxy, like most of the observed stars; this would probably not solve Ostriker's stability problem. Moreover, Oort included the undetected component of the interstellar medium, dust and gas, as part of the dark component so that it did not seem, at the time, particularly mysterious.

But observational evidence in support of the idea that spiral galaxies possessed a substantial, more extended unseen component was beginning to appear in the early 1970s. My first real position, in 1972, was at the National Radio Astronomy Observatory (NRAO) in Charlottesville, Virginia. This was primarily an observational institute and I was known as a "house theoretician". Radio astronomers at NRAO, such as Mort Roberts and Seth Shostak, had been observing the distribution and motion of neutral hydrogen in the outer parts of galaxies through the spectral line emitted by hydrogen at a wavelength of 21 cm (Chapter 4). They noticed that the rotational velocity of the gas does not seem to be declining with distance from the centers of galaxies as it should for a bounded mass distribution. The rotation velocity appeared to be constant well beyond the visible image of the galaxy. This was a very contentious result at the time, with heated debates about telescope side lobes and possible warping of the gas layers in spiral galaxies, but it was a clear early indication that there is a real discrepancy between the dynamical and visible mass in galaxies. And it was in complete accordance with the suggestion of Ostriker and Peebles.

Later in my career, in 1977, I accepted a position at the Kapteyn Astronomical Institute at the University of Groningen in the Netherlands, again, as a house theoretician at a primarily observational institute. A few years before that, the synthesis radio telescope at Westerbork, a one and one-half kilometer array of dishes used as a single telescope, had begun operating and was being applied to observe the distribution and motion of neutral hydrogen in spiral galaxies with relatively high spatial and velocity resolution. The radio astronomers at Groningen were making precise measurements of the "rotation curves" of spiral galaxies – how the gas rotates as a function of distance from the center well beyond the visible object. Consistent with the earlier observations, the rotation velocity was not seen to decline but remained constant with distance implying that the gas, although well beyond most of the light of the galaxies, is still immersed in the mass distribution of the galaxy – that the mass in the outer regions of the galaxies is dark. Coming from Princeton and from NRAO, with all my theoretical and observational prejudices, this was not a surprising result for me. I realize now that I was not as excited as I should have been. Westerbork was producing the most convincing and direct observational confirmation of an idea that was still quite tentative – the idea that the visible parts of galaxies were a tiny, shiny central component of a vast dark system.

Evidence from other sources had been mounting as well. High-resolution measurements of rotation curves from spectroscopic observations of optical emission lines by Vera Rubin and her collaborators were beginning to appear in the literature – these rotation curves were also flat out to the optical edges of the spiral galaxies. Because the rotation velocity was not measured beyond the optical image, this did not constitute compelling evidence for dark matter, as I will discuss in Chapter 5; but that was not the perception at the time. These observations had an enormous impact on the growing realization that there was a substantial dark matter component in spiral galaxies. By the early 1980s this viewpoint was rapidly becoming the paradigm.

My own interest has been mostly centered on galaxies and the manifestations of the mass discrepancy on a galaxy scale. But evidence was mounting on other scales as well. In the 1970s satellites that could observe the sky at X-ray wavelengths (this radiation does not penetrate the atmosphere of the Earth) were launched into Earth orbit. It was discovered that distant clusters of galaxies were powerful sources of X-rays and that this emission is thermal radiation from vast pools of hot gas filling the clusters. In fact, the mass of gas generally exceeds that of the stars in galaxies by a factor of two or three. Could this be Zwicky's missing cluster mass? For such a gaseous object in equilibrium one can, by measuring the temperature and density distribution of the gas, determine the gravitational field and, hence, with Newton's law of gravity, the mass of the entire system. When this was done, it became apparent that most of the mass of clusters of galaxies was still unseen; that the clusters contained at least five or six times more mass than was detected in stars and gas. Was this dark matter the same as that in individual galaxies? It was, and is, generally assumed to be so.

It was also becoming evident in the late 1970s that something is missing on a cosmological scale. The Universe is typically modeled as an expanding, isotropic, homogeneous fluid, and certainly on the largest scales it appears to be that way. The cosmic microwave background radiation, (the CMB) discovered in 1965 by Arno Penzias and Robert Wilson, should reflect density fluctuations in the cosmic fluid when the Universe was only 300 000 years old – when protons and electrons combined to make neutral hydrogen and the radiation decoupled from the matter. These fluctuations in the CMB were looked for and not found at the level of about one part in 10 000. This means that all of the structure that we observe in the Universe – from stars to galaxies to clusters of galaxies and to super clusters – has formed in the last 14 billion years or so by the gravitational growth of incredibly small fluctuations. This just did not seem possible in the context of the standard theory of gravitational instability. A solution to this problem is to add dark matter, but a special kind of dark matter: matter consisting of particles that interacts with light or ordinary (baryonic) matter primarily through gravity – "non-baryonic" dark matter. Because it is decoupled from the radiation, this dark matter fluid can begin to gravitationally collapse sooner than the normal baryonic matter – before the recombination of hydrogen. This gives the observed structure time to form from the very small density fluctuations. So dark matter on a cosmological scale appeared to be a necessity as well. (The missing fluctuations were finally seen at a level of 10^{-5} by the COBE satellite in 1992. See Smoot *et al.*, 1992.)

But a completely new aspect of the dark matter problem emerged from these cosmological considerations. This cosmological dark matter is very different than what had originally been imagined for the dark halos surrounding galaxies. It is not small or dead stars, but subatomic particles – and not the ordinary subatomic particles like protons and neutrons, but something else which interacts very weakly – neutrinos perhaps, or something even more exotic, something not yet detected in terrestrial laboratories. At about the same time, particle physics theory was advancing beyond its so-called standard model. New ideas on the unification of forces were being proposed – grand unification and then, supersymmetry. These new theories provide a host of particle dark matter candidates in addition to the modest neutrino. Subatomic particles possess an attribute called "spin" that is quantized (it comes in distinct lumps). In supersymmetry every known standard-model particle is required to have a supersymmetric partner that differs by half-integer spin. So this theory, in effect, doubles the number of possible particles. Only one of these hypothetical particles – the lowest mass superpartner – is stable and long-lived and could be the dark matter. But because of this possibility, physicists became very excited about the prospect of dark matter – some even appeared to believe that they had invented dark matter. This union of astronomers, cosmologists and particle physicists led to the development of a new, interdisciplinary subject – astroparticle physics. Once again, astronomical observations had spawned not only a new paradigm, but a new field of study.

In the spring of 1982, I was taking a four-month sabbatical at NRAO and enjoying the Virginia spring while working on an absolutely unrelated topic – the jets observed to be emanating from some active galactic nuclei. In those days, preprints of scientific articles – pre-publication versions of papers which were usually in press already – were not placed on the Internet – there was no Internet – but were distributed in printed form between various scientific institutes. NRAO was definitely on this preprint circuit, and at some point, around April 1982, three preprints arrived from the Institute of Advanced Study in Princeton. These were preprints on the missing matter problem authored by an Israeli physicist, Mordehai Milgrom. I had actually encountered Milgrom before in a rather competitive way; he had independently developed a model that I had proposed some years before – a model for compact radio sources with apparent faster-than-light motion. But here, in these articles, Milgrom was proposing a very radical new idea – and not one that I could claim to have thought of. He was suggesting there is no dark matter but that the usual Newtonian dynamics or gravity was not applicable on these extragalactic scales. His hypothesis was called "modified Newtonian dynamics" or MOND for short. These preprints first brought home to me the realization that, after all, dark matter is a sort of ether – a medium that is necessary to make observations consistent with the expectations of existing theory. If the theory is inappropriate on these scales, then perhaps there is no ether.

Now Milgrom's idea is basically very simple: Newtonian dynamics is modified at low accelerations – that the familiar old formula $F = ma$ becomes more like $F = ma^2/a_0$ at accelerations below a critical value a_0. This simple modification appears to accomplish a great deal. It yields flat galaxy rotation curves in the limit of large radius (low acceleration), and provides a relation between the mass of a galaxy and its rotation velocity, or if mass is proportional to luminosity, a luminosity–rotation velocity relation. In fact, such a relation had been observed years before by Brent Tully and Rick Fisher – the Tully–Fisher relation – and Milgrom's acceleration-based modification provided a simple explanation of this correlation as resulting from existent physical law, as opposed to dark matter which attributed such scaling relations to the contingencies of galaxy formation. Moreover, MOND predicts that high-surface-brightness systems, like globular star clusters for example, should have no apparent dark matter problem within the visible object, and that low-surface-brightness systems, such as the dwarf spheroidal satellites of our own Galaxy, should have a large discrepancy.

I was fascinated by this idea, but I thought that it was probably not correct. Such a drastic modification would surely have other consequences – consequences for cosmology and large-scale structure in the Universe. It seemed to me that it was not just sufficient to explain a few facts about galaxies, the idea had to fit into a much larger picture. There is much more to explain than galaxies.

I let this go for a while, but then, a couple of years later, back in Groningen, I had my own idea. I read a paper by a French physicist, Joel Sherck, who proposed that, consistent with supersymmetry or its follower, supergravity, additional fields might exist in the Universe; fields which couple to matter with gravitational strength. One possibility is a vector field, but vector fields, like electromagnetism, produce a repulsive force between similar particles – an anti-gravity. The force would be carried by a particle, a so-called vector boson. Sherck wanted this vector boson to have a finite mass and therefore a limited range, but a range so small that it would have no actual macroscopic effect on scales of one meter or so where the inverse-square law of gravity had been carefully measured (the larger the mass of the field, the smaller its range). I picked up on this suggestion and warped it to my own purpose.

How could a repulsive force yield flat rotation curves? I thought – perhaps gravity, locally, is a mixture of repulsion and attraction, but slightly more attraction. Suppose also that the vector boson which mediates the repulsive force has such a small mass that its range would be on the scale of galaxies? This would mean that on a scale larger than a galaxy the repulsive force would die away leaving pure attraction. It would be possible to have a larger effective gravitational attraction on extragalactic scales than on the sub-galactic scale. Adjusting the mass of the vector boson correctly and the ratio of repulsion to attraction correctly, one could produce flat rotation curves for spiral galaxies over a range of about a factor of 10 in radius. This, I thought, led to a more cosmologically acceptable model, because on the largest scale, there was a return to inverse-square attraction, and the Universe behaved as it would in the standard picture with 10 times more dark than visible matter. I might add here that I didn't know very much about general relativity in those days and didn't realize that my proposal would violate the local universality of free fall (first tested by Galileo in his famous, but probably fictional, Tower of Pisa experiment) in a very blatant and detectable way.

I immediately submitted a short paper to *Astronomy and Astrophysics* (the European journal) and waited to see what would happen. There were two reviewers of the paper, one of whom was Milgrom. He was very negative in his report. He pointed out that such a modification would, indeed, lead to a Tully–Fisher law, but the wrong Tully–Fisher law: $L \propto V^2$ instead of $L \propto V^4$, as is, so he claimed, more consistent with observations. I protested. I thought that the form of the Tully–Fisher law was not so evident at that point; it seems to depend upon the color in which the luminosity is measured, and in blue light it is more like $L \propto V^2$. I was so attracted by my idea that I thought that it must be published, and after much pleading with the editor (who occupied an office a few doors from my own), it was.

I cherished this idea for several years more, but then, the reality of galaxy phenomenology caught up with me in the form of two facts. The first fact is that Milgrom was right about the form of the Tully–Fisher law – when measured in the near-infrared emission from stars (the radiation from the old, low-mass stars that are the dominant component of the stellar disk), the relation really is more like $L \propto V^4$, as he said. The second is this: larger galaxies do not exhibit a larger discrepancy – big galaxies do not need more "dark matter". I had proposed a modification of gravity attached to a definite fixed-length scale. This means that galaxies which are larger than this length scale should have a larger discrepancy and smaller galaxies a smaller discrepancy or even no discrepancy at all. Being at an institute that was primarily observational and producing new rotation curves every day, I realized that this was not true. There are very small galaxies with a large discrepancy, and very large galaxies with a small discrepancy. The discrepancy seems to be more dependent upon surface brightness (the energy of radiation emerging per second per square meter at the source) than size, and surface brightness, in so far as it reflects surface density, is proportional to acceleration.

My idea seemed pitiful and lonely without any observational support, so even I had to abandon it. I think, actually, that many scientists have trouble with this. We become too deeply attached to ideas because they are ours – but confronted by the facts, painful though it is, we are forced to forsake our pet theories. It must have been around 1985 when I realized that Milgrom was right. The only sort of modification of gravity or dynamics that could possibly replace dark matter was a modification attached to an acceleration scale. Then began for me a long period, still continuing, of work on MOND – observational and theoretical. I corresponded with and met another Israeli colleague of Milgrom's – the physicist Jacob Bekenstein. Jacob was a relativist – an expert in general relativity well known for his work on black holes – and he believed that MOND should be viewed as a modification of the theory of gravity. Jacob thought, and I agreed, that if MOND is to ever be acceptable it must connect to more familiar physics – it must be an aspect of a more general theory of gravity or inertia. I still think that this is true, but it is also true that what is "familiar" changes as well.

But what of dark matter? If MOND is right, is dark matter wrong? Simply defined, MOND is an algorithm for calculating the gravitational force in an astronomical object, from the observed distribution of ordinary baryonic (detectable) matter. And it works – at least on the scale of galaxies. Because it works, this is very problematic for dark matter – at least on the scale of galaxies. It would seem to imply a very precise coupling between dark matter and baryonic matter – a coupling that is not comprehensible in the context of standard or "cold" dark matter. On the other hand, cold dark matter is quite successful on cosmological scales; it predicts the formation of observed large-scale structure and the magnitude and distribution of fluctuations in the primordial cosmic microwave background. How could these two be reconciled?

But another interesting twist, which no one really imagined 20 years ago, emerged in the late 1990s: dark matter alone is not sufficient; it appears that, on a cosmological scale, "dark energy" is also required. This is a mysterious fluid with a negative pressure that does not dilute as the Universe expands and leads to the accelerated expansion of the Universe. In Einstein's theory of gravity, general relativity, the dark energy is embodied by the so-called cosmological constant. It may also be identified with the energy density of the vacuum, a concept of modern quantum field theory in which "empty" space is actually filled with virtual particles popping into and out of existence – virtual but gravitating. In this case, the vacuum energy density should be many orders of magnitude larger than it is observed to be; in fact, so large that the Universe as we observe it would be impossible. The observation of a tiny value for the vacuum energy density, tiny in terms of the expectations of quantum field theory, is one of the greatest puzzles in modern physics.

Now that we "know" the composition of the Universe, some cosmologists have become quite triumphal. There certainly has been enormous progress, but given this very strange composition – a mysterious and unnatural dark energy as well as a dark matter fluid which has not been detected by any means other than its gravitational influence – triumphalism seems to be premature. To me, it appears presumptuous to assume that we human beings at this point in our development understand either the material content of the Universe or all of its physical laws.

Here I want to describe the process of discovery over the past 40 years that has led to the development of the dark matter paradigm as well as the now standard cosmological model. Of course, these developments have spawned not only the paradigm but also its alternative, as I will discuss in Chapter 10. I will discuss the dark matter vs. MOND controversy as a conflict of paradigms, but my primary purpose is to provide the reader with a reasonably objective view of the major developments in the emergence of the dark matter–dark energy view of the world. Most of my own experience is in the field of galactic astronomy. So in this discussion I will emphasize galaxy-scale phenomenology which provides, after all, the primary observational evidence for dark matter that clusters on a small scale and is, possibly, directly detectable locally.

I will not discuss one very interesting aspect of the dark matter problem: the development of the astronomy of gravitational micro-lensing with the goal of detecting "massive compact halo objects" or MACHOs. This was a brilliant observational technique that spawned a new arena of astronomical research and provided the direct observational evidence that normal "baryonic" matter in the form of stellar and sub-stellar mass objects could not be the principal constituent of dark matter halos about galaxies. I refer the reader to the book on dark matter by Freeman and McNamara (2006) for a highly readable account of this development.

The level of this discussion should be appropriate for professionals as well as beginning students and interested readers with some scientific background. Therefore, the presentation is essentially non-mathematical. However, I include a pedagogic appendix that is primarily for those who are less familiar with astronomical concepts and terminology. Here I provide the most relevant formulae and definitions. This can safely be skipped by professionals or more advanced students, but the scientifically literate reader may find this survey to be useful as an introduction to the jargon as well as the more quantitative aspects of the problem. In particular, I focus on the following points:

(1) Electromagnetic radiation is the primary (but not the only) medium for observing objects in the distant Universe. What is the nature of electromagnetic radiation? How is it emitted and how does it propagate? What are spectral lines and how are they formed? How can we measure the velocity of an astronomical object toward or away from us by using spectral lines?

(2) It is important to be acquainted with aspects of scale in astronomy. What are the units of distance appropriate to galactic and extragalactic problems? How do we measure distance? What do we mean by apparent brightness and intrinsic luminosity of a star

or galaxy, and what are the appropriate physical units? What is meant by the surface brightness of astronomical objects? How do we measure the color and composition of stars and galaxies? What are the characteristics and morphological types of galaxies? What is the mass scale, luminosity, star and gas content of extragalactic objects?

(3) Familiarity with a few basic physical concepts is necessary – Newton's laws and classical mechanics – because this is how we measure the mass of gravitating systems.

(4) Dark matter is thought to be a substantial component of the entire Universe and required for the formation of observed structures such as galaxies and clusters; therefore I consider a few basic concepts of cosmology, which is the study of the structure and evolution of the Universe as a whole. I define the fundamental density parameter of cosmology and describe the known constituents of the Universe – visible matter and electromagnetic radiation. What is baryonic or non-baryonic matter? What do we mean by "dark energy"? I discuss the thermal history of the Universe and take up the question of how structure – stars, galaxies, clusters of galaxies – can form from an originally hot, highly homogeneous expanding Universe.

I assume throughout that the reader is familiar with scientific notation; that is, instead of writing 1 000 000 000, I write 10^9, or 10^{-3} instead of 0.001. In the text, I write only the most basic equations, often without derivation, because of my generally qualitative and historical approach to this subject.

The style of this discussion is essentially narrative and personal. I have not only witnessed, but in some cases, been involved in these developments, so I do have a very direct interest. I have been privileged to work at institutes where much of the initial work on the dark matter problem, especially with respect to galaxies, has been carried out, and I know a number of the principal players who have shared their thoughts and enthusiasm. I have learned a great deal from the dark matter problem, not only about dark matter but also about the way in which science progresses and how scientists work. I will conclude with some general remarks on these sociological aspects of science as exemplified by the dark matter problem.

This work is a personal and by no means a complete or encyclopedic history of the subject. So I will not cite everyone who has made significant contributions to the study of the dark matter problem; I apologize in advance to those who may feel slighted. I do think that I have included reference to most of the major contributors in this field.

Finally, I hope that I can convey to the more general reader a sense of the excitement in this ongoing adventure of discovery and at least make the case that the adventure is far from complete.

2

Early history of the dark matter hypothesis

2.1 Prehistory

Dark matter, in an astronomical sense, is introduced to explain the difference between how objects in the sky ought to move, according to some preconceived notion, and how they are actually observed to move. In that sense, the first application of this concept was employed, as one might expect, by the Greeks. In the fourth century BC Eudoxus of Cnidus, a student of Plato, proposed that the stars and the Sun were attached to transparent spheres with the Earth at the center. The spheres rotated about the Earth: the outer "stellar" sphere once per day, and the inner "solar" sphere also once per day with an additional annual motion with respect to the stellar sphere. The spheres were not actually dark but transparent, so that we could see the stellar sphere through the solar sphere. The spheres did not possess the attribute of mass – or weight – so perhaps it is not proper to speak of them as matter. But they were certainly unseen constructs proposed to explain the observations. The most famous student of Plato, Aristotle, increased the number of spheres to 55 to account for planetary motion, and he seemed to believe in the actual reality of these spheres.

The system was codified by Ptolemy in the second century AD and reigned supreme for about 1300 years. By the time of Copernicus the entire construction had become very complex with spheres on spheres and off-center spheres in order to explain the increasingly precise observations of planetary motions. In this sense the celestial spheres began to adumbrate the modern concept of dark matter: the number of spheres and their attributes were enlarged to account for more detailed observations. It worked perfectly, but the system finally collapsed as a result of its complexity and the emergence of the simpler Copernican system – verified by the telescopic observations of Galileo. This of course was the beginning of modern science and initiated a development culminating with Newton's laws of motion and gravitational attraction.

It can be argued that the first success of the dark matter hypothesis, in the context of Newtonian gravity, is due to the French mathematician Urbain Le Verrier who, puzzled by small systematic peculiarities in the orbital motion of Uranus in the context of Newtonian gravity, pointed out that the existence of an unseen planet beyond the orbit of Uranus would resolve the anomalies. He did more than that: he actually determined the orbit of the undiscovered planet and, in 1846, predicted its position in the sky. Neptune was then quickly discovered by Johann Galle at the Berlin observatory (Le Verrier was independently followed in his prediction a few days later by the Englishman, John Adams).

These developments must actually be viewed as an outstanding success of Newtonian theory, which we now know works extremely well on the scale of the Solar System apart from tiny corrections due to the more complete theory of general relativity. But the success was realized only with the independent (i.e., visual) detection of Neptune; if no planet had been seen, Le Verrier's prediction would have remained an unconfirmed hypothesis. In that case, the anomalous motion of Uranus would have pointed to a breakdown of Newtonian gravity in the Solar System. This has relevance to the modern concept of dark matter, as we shall see.

2.2 Zwicky and the modern concept of dark matter

The reconciliation of astronomical observations with Newtonian dynamics is also the original motivation for the modern hypothesis of dark matter. But, as it has developed, this new form of dark matter is perceived to be a pervasive fluid filling the Universe and comprising the dominant component of bound astronomical systems like galaxies or clusters of galaxies, detectable only by its gravitational influence in these systems. Unquestionably, Fritz Zwicky was the first to propose this form of dark matter, although for 30 years after his proposal, it was largely unappreciated; astronomers did not see Zwicky's anomaly as a crisis leading to a possible paradigm shift. It took about 40 years for Zwicky's insight to be fully accepted.

Fritz Zwicky was one of those rare unorthodox geniuses who occasionally emerge in astronomy or, for that matter, in any field. A Swiss citizen who lived and worked in the United States (Caltech) for many years, Zwicky made profound contributions to modern astronomy and astrophysics – from the observations and theory of exploding stars (supernovae) and their remnants (neutron stars) to the study and classification of galaxies and clusters of galaxies. He was decades ahead of his time, and the fact that much of his work was ignored by contemporaries no doubt contributed to his famously irascible behavior; he was, by any criterion, a difficult person.

Fig. 2.1. The Coma cluster of galaxies. This is a highly regular gravitationally bound system of thousands of galaxies at a distance of about 100 Mpc (NASA, SDSS).

In 1933, Zwicky looked at radial velocity measurements (the component of velocity along the line-of-sight) of several individual galaxies in the well-known Coma cluster of galaxies (Fig. 2.1) and he noticed something quite striking: the galaxies seemed to be moving too fast for the amount of visible matter in the cluster. If one just adds up the mass in the cluster by assuming that every galaxy has a mass-to-light ratio of about one in solar units, then the individual galaxies are moving so fast that they should quickly (by astronomical standards) escape; in other words, we should not observe a cluster at all because it would have long since dispersed. He published the results of his analysis in German in the Swiss journal *Acta Helvetica Physica*.

In what was then an original application of classical mechanics to extragalactic astronomy, Zwicky used the "virial theorem" discussed in the Appendix to estimate the dynamical mass of the Coma cluster (see eq. A4.12). The mass M of a self-gravitating system in equilibrium (not changing its form) is roughly

$$M = RV^2/G \qquad (2.1)$$

where R is the characteristic radius of the system and V is the random velocity of the galaxies in the system (the velocity spread) and G is Newton's gravitational constant. To get R we need to know the distance to the cluster, but, because of the recently discovered Hubble law (Section A2), and having the average recession velocity of the cluster galaxies, Zwicky could estimate that the distance was about 50 Mpc (the modern estimate is twice this because of improved determination of the Hubble constant). It turned out that R was about 1 Mpc. Zwicky had the measured radial velocities of only eight member galaxies out of roughly 1000 in the entire cluster and was immediately struck by the fact that these radial velocities ranged over 1000 km/s. He took this as an estimate in the spread of galaxy velocities in the cluster, and applying the virial theorem immediately determined a huge mass of 3×10^{14} solar masses ($M_\odot$). But, only about 10^{12} $M_\odot$ could be accounted for by the visible galaxies. The cluster appeared to contain several hundred times more mass than that of stars in the visible galaxies. Thus, Zwicky summarized "Should this turn out to be true, the surprising result would follow that *dark matter* is present in a much higher density than radiating matter" (my italics). Hence, the first appearance of the term "dark matter" in the context of extragalactic astronomy.

Three years later, one of Zwicky's colleagues at the Mount Wilson Observatory, Sinclair Smith (1936), repeated the analysis for the relatively nearby Virgo cluster of galaxies. Smith had at hand radial-velocity measurements of 30 member galaxies and came to roughly the same conclusion – the implied mass-to-light ratio was more than 100 in solar units. Zwicky, revisiting this problem in 1937 in an article in the more widely read *Astrophysical Journal* was somewhat dismissive of the Virgo result: he argued that Virgo was a large irregular and diffuse cluster – unlike the richer and more symmetric Coma – and therefore, application of the virial theorem is more questionable. He reconsidered the Coma cluster and came essentially to the same conclusion: the mass-to-light ratio in Coma must be in excess of 500 – several hundred times larger than in the solar neighborhood of the Milky Way. He considered the possibility that the cluster may not be in equilibrium – that it may not obey the virial theorem. Then there are two possibilities: either the system is dominated by kinetic energy, in which case it should fly apart in a few billion years. Then, of course, the question is why we observe such rich clusters of galaxies at all. Or the system is dominated by its gravitational potential energy, in which case there is even more unseen mass.

With respect to the unseen mass, Zwicky considered two alternatives. Either the individual galaxies are much more massive than would be suggested by their luminosity (i.e., a much larger mass-to-light ratio than locally). Or there is undetected "inter-nebular matter" ("intergalactic matter" in the current terminology). He raised another possibility: "It should also be noticed that the Virial Theorem as applied to clusters of nebulae provides for a test of the validity of the inverse square law of gravitational forces. This is of fundamental interest because of the enormous

distances which separate the gravitating bodies whose motions are investigated."
In other words, this could provide a test of Newtonian gravity on the largest possible scale. He no doubt realized, but did not state, that such a test would only
be possible if the unseen matter could, in fact, be independently detected. In the
absence of an alternative theory, judgements on the validity of Newtonian gravity
are only valid if the total mass distribution can be observed independently of its
gravitational influence.

2.3 Dark matter on galaxy scales

In 1932, one year before Zwicky's paper on the Coma cluster, the Dutch
astronomer, Jan Oort, first used the term "dark matter" in discussing the mass density in the disk of the Milky Way as deduced from the motion of stars above the
galactic plane. From the kinematic data he calculated that only about one-third of
the dynamically inferred mass was present in bright visible stars. It is clear from the
context that, in characterizing the remainder as dark, Oort was describing all matter
not in the form of visible stars with luminosity comparable to or larger than that
of the Sun. Oort included both low-mass stars and the undetected component of
the interstellar medium (the gas and dust between the stars) in the category of dark
matter, and concluded that when a proper accounting of all forms of conventional
undetected matter is included, the resulting mass is not likely to be inconsistent
with the dynamical mass. The first suggestion of a significant dynamical discrepancy on the scale of galaxies emerged a few years later from observations of
rotation curves.

The term "rotation curve" is very important for the rest of my discussion, so I
should carefully define it: the rotation curve of a galaxy is the rotational velocity
about the center, given in km/s, plotted as a function of radius usually given in kpc
(although sometimes the radius is given in angular measure, such as arc seconds).
The rotation curve may be measured if there are spectral emission lines from hot
gaseous regions throughout the galaxy; it is also possible, but more difficult to
measure the rotation curves in absorption lines from the stars making up the disk
of the galaxy, but gas is thought to be a better tracer of the true circular velocity in
galaxies because the stars have a higher random velocity (i.e., they move on more
elliptical orbits). We will see below that rotation curves may also be measured
using the 21-cm line of neutral hydrogen – a line at radio wavelengths – but this
became possible only in the 1950s. The rotation curve, if it results from true circular
motion, is a tracer of the force, and hence, the mass distribution in the galaxy (see
Section A4).

In his paper of 1937 Zwicky considered and dismissed the possibility of deriv-
ing the mass distribution of individual galaxies from the internal motion of the

stars – the random velocity or the rotation curve. He thought, erroneously, that the gravitational attraction between pairs of stars would add an effective viscosity to the stellar system, making the rotation law an unreliable probe of the gravitational force. This misconception was corrected a few years later (1941) by Subrahmanyan Chandrasekhar, who demonstrated that gravitational interactions between pairs of individual stars were absolutely negligible in a rich stellar system such as a galaxy.

A measurement of the rotation curve of M31, the large, relatively nearby Andromeda galaxy, was reported in 1939 by Horace Babcock in his PhD dissertation at the University of California in Berkeley. A photographic image of this large spectacular spiral galaxy is shown in Fig. 2.2. Babcock determined the rotation curve by measuring the radial velocities of optical emission-line regions extending out to 100 arc minutes (about 20 kpc) from the center of Andromeda (see the Appendix for a discussion of the Doppler shift and velocity measurements). Quite remarkably, there was no indication of a Keplerian decline in the rotation velocity, as one would predict from the light distribution. Babcock concluded that the mass-to-light ratio must systematically increase with radius, from 20 to more than 60. He commented that "... the great range in the calculated ratio of mass to luminosity in proceeding outward from the nucleus suggests that absorption (of light) plays a very important role in the outer portions of the spiral, or, perhaps, that new

Fig. 2.2. The great spiral galaxy in Andromeda. At a distance of about 700 kpc, this is the nearest large spiral galaxy to our own Milky Way. Notice the small spheroidal companion above the plane of M31. This is M32; Schwarzschild assumed that M32 was distorting (warping) the plane of M31 and used the evident warp to estimate the mass of this companion.

dynamical considerations are required which will permit of a small relative mass in the outer parts." By this last comment he meant a modification of Newtonian dynamics. In other words, Babcock considered two possible explanations: either there was more absorption of the starlight in the outer regions (by dust), or Newton's laws did not apply on these scales; he did not consider dark matter as such. Comparing Babcock's rotation curve with more modern results we see that his conclusion was overstated. There is some evidence in Babcock's observations for a constant rotation velocity, but, as became evident later, this in itself does not constitute evidence for an increasing mass-to-light ratio in the outer regions.

This question seems to have vanished as a significant scientific issue for the following 15 years. Of course, the Second World War intervened, and many scientists who normally spent their time thinking about such abstract problems became otherwise employed. One of these was Martin Schwarzschild, the son of the famous Karl Schwarzschild who wrote down the first mathematical solution to Einstein's equation for the gravitational field about a spherical object. Although Karl had perished while serving in the German army during the First World War, Martin was forced to flee from the Nazi regime, arriving, after a stay in Norway, in the United States where he worked first at Harvard and then at Columbia. On the day after Pearl Harbor, he offered his services as a private soldier in the US Army (later, in Italy, having advanced to Army Intelligence, he had no end of problems because of his heavy German accent).

After the war, Schwarzschild returned to astronomy and finally to Princeton University Observatory, where he carried out fundamental theoretical work on the structure and evolution of stars. But in 1954, 20 years after Zwicky's original discussion of the extraordinary mass-to-light ratio in the Coma cluster, he revisited the issue of the mass-to-light ratio in individual galaxies. In a remarkable paper that was more than two decades ahead of its time (both with respect to the issue considered and in the technique employed), Schwarzschild addressed the very modern question of whether or not the mass and light in galaxies has the same distribution. Is there a constant mass-to-light ratio in galaxies, or, as claimed by Babcock, does the mass-to-light ratio increase moving radially outward? Schwarzschild used more recent spectroscopic observations of M31 by Nicholas Mayall (1951) at Mount Wilson and constructed a rotation curve (Fig. 2.3) which shows considerable scatter. But he notes that, contrary to early observations (i.e., Babcock's), there is no indication of solid-body rotation in the inner regions, but that the plotted points "suggest a fairly constant circular velocity over the whole interval from 25' to 115'" (6 to 26 kpc). He notes that these new observations lack sufficient accuracy to define the mass distribution in the galaxy, but he turns the problem around. Are the observations consistent with a mass distribution which follows the light distribution?

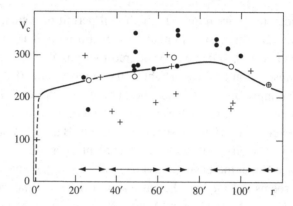

Fig. 2.3. The solid points and crosses show the rotation curve of M31 as determined by optical emission-line measurements by Mayall. The open points show the average values of the observed rotation curve (averaged over intervals in radius shown by the arrows). The solid curve is Schwarzschild's theoretical curve calculated from the observed distribution of light assuming a constant mass-to-light ratio and that the stellar mass is distributed in a thin disk. Here the radii are given in units of minutes of arc and the velocity, as usual, in km/s. (From Schwarzschild, 1954.)

To answer this question he assumed that the light exactly traced the mass, and that the mass was distributed in a thin disk. He then applied Newton's law (which allows one to calculate the force distribution from the mass distribution) to determine the radial force distribution and, hence, the rotation curve. This procedure sets the shape of the rotation curve, but of course, there is one more free parameter in all of this – the parameter that determines the amplitude or height of the rotation curve – and that is the mass-to-light ratio. By adjusting the mass-to-light ratio he found that he could quite nicely match the shape and amplitude of the observed rotation curve, certainly within the errors of the observations (see Fig. 2.3). In other words, there was absolutely no evidence that the mass-to-light ratio increased in the outer regions of M31. He did derive the improbably high mass-to-light ratio of 16 (partly due to the fact the distance to M31 was assumed to be about one-half of its presently accepted distance), but the fact remains that the observations do not require this quantity to vary. The significant implication of Schwarzschild's paper is that a constant or flat rotation curve measured within the optical disk does not, in itself, require an increasing mass-to-light ratio in the outer regions of a disk, even though the intensity of light is falling off (much more on this point follows later).

Schwarzschild went on to consider mass-to-light ratios in elliptical galaxies – the spheroidal gas-free systems found primarily in groups and clusters. Based upon an early determination of the random velocities of stars in one such system, NGC 3115 (the "spindle galaxy" which, we know now, is actually not an elliptical galaxy but a disk galaxy with a ring of gas around its polar axis), he concluded that the mass-to-light ratio was exceedingly high: about 100 (this high value was due to the imprecision of the kinematic observations). He went on to consider M32, the small elliptical companion of M31. Here he noted that M31 appears to be asymmetric and disturbed, and he assumed that the asymmetry was due to the perturbing influence of M32. From this he estimated a mass (2.5×10^{10} $M_\odot$) and determined an M/L of 200 (we now know from the internal kinematics of stars in this system that this is a considerable overestimate, at least for the inner regions where the stars are found).

Schwarzschild attributed the difference between ellipticals and spirals to differences in stellar populations: as discussed in the Appendix, spirals contain a large fraction of young bright stars with a very low mass-to-light ratio, whereas ellipticals are composed primarily of old low-mass, dim stars. Although not so much was known about the mass-to-light ratios of stellar populations then, Schwarzschild did realize that the mass distribution of stars, such as those found in the solar neighborhood, could not possibly exhibit such a large M/L, and he speculated that there could be many dead stars – cool white dwarfs – which had exhausted their fuel and contributed mass but no light.

If the mass-to-light ratio in individual elliptical galaxies is 200 then Zwicky's conclusion on the Coma cluster and the need for dark matter is not so remarkable after all. Schwarzschild considered Coma and estimated an M/L of about 800. So the amount of missing mass outside of galaxies may be only a factor of four. Schwarzschild concludes that this "bewilderingly high value for the mass–luminosity ratio (of Coma) must be considered as very uncertain since the mass and particularly the luminosity of the Coma cluster are still poorly determined".

And that is essentially where the issue stood for the next 10 to 15 years. Most astronomers felt that Zwicky's discrepancy in clusters of galaxies was not a real discrepancy – that the issue would be resolved when a more exact accounting of the mass and light in clusters could be made. In 1961 a conference was held at Santa Barbara on the stability of clusters and groups of galaxies. The essential question was whether or not it was appropriate to apply the virial theorem to clusters. The Armenian astronomer, Viktor Ambartsumian, had quite radical ideas on galaxy formation: he thought that galaxies might be ejected from other galaxies via some, as yet, undetermined physical process. If so, then a grouping of galaxies on the sky might not be a gravitationally bound system but might be due to newly born galaxies still in the neighborhood and moving rapidly away from a parent. The conclusion of most participants was that such a mechanism, if it existed in reality, could not apply to the great regular clusters of galaxies such as Coma. The problem of the large implied mass-to-light ratio remained, although, again, the general feeling was that the discrepancy would disappear with improving observations or theoretical understanding.

In the meantime, however, other developments were proceeding in the study of galaxy kinematics – great improvements in the accuracy and sensitivity of spectroscopic observations and the emergence of a completely new tool in observational astronomy – radio astronomy and the 21-cm line of neutral hydrogen.

2.4 Radio astronomy: a new tool for galactic astronomy

In astronomy, the opening of a new window, for example, a new wavelength regime of the electromagnetic spectrum, has most often led to significant discoveries and even, occasionally, to a fundamental paradigm shift. So the discovery by Carl Jansky in 1933 of extraterrestrial radio waves has had a major impact on our view of the Universe, and has added quasars and pulsars and the cosmic microwave background to the list of previously unknown, and barely imagined, extraterrestrial exotica.

In 1940 Jan Oort realized that a spectral line in the radio regime of the electromagnetic spectrum would be extremely useful in the study of galactic structure. This is because, unlike visible starlight, radio waves are not dimmed by absorption due to interstellar dust. Oort asked his Leiden student Henk van de Hulst to look into the problem: are there spectral lines emitted by interstellar material at centimeter wavelengths? Van de Hulst discovered that there is indeed such a line – a line emitted at 21 cm by neutral hydrogen. Hydrogen is the dominant element in the Universe and is the simplest atom, consisting of a single electron and a single proton. The 21-cm line originates from a so-called hyperfine transition in the hydrogen atom: when the electron spin is parallel to the proton spin, the energy is slightly higher than when they are anti-aligned. A transition between these two states gives rise to the spectral line. Now this is very useful because the interstellar medium consists largely of neutral hydrogen: if the line could be detected then the distribution and motion of hydrogen could be mapped at vast distances across the Milky Way Galaxy, and, with sufficient sensitivity and resolution, in other spiral galaxies. The line was actually detected in 1951 by the Americans, Harold Ewen and Edward Purcell, and confirmed shortly afterwards by Lex Muller and Jan Oort (1951).

Jan Oort was a great organizer and manager. But his organizational skill was entirely driven by scientific considerations, and his scientific judgement was superb. After the war the Dutch countryside was littered with radar dishes left behind by the occupying German forces, and Oort and his colleagues saw that these could be used to establish a program of radio astronomy. For Oort the primary scientific objective was to derive the structure and dynamics of the Galaxy, and he saw that an essential tool for this is the 21-cm line of neutral hydrogen. The principal radio observatory was constructed at Dwingeloo in the relatively sparsely

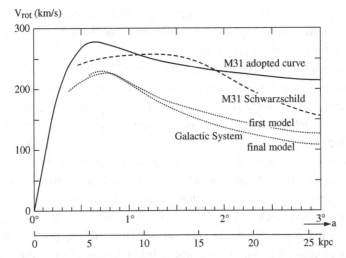

Fig. 2.4. The solid curve is the model rotation curve of M31 fitted to the Dwingeloo 21-cm line observations by van de Hulst *et al.* in 1957. The dashed curve is the model rotation curve of M31 calculated by Schwarzschild in 1954 assuming that mass traces light. Within the optical disk, these two agree to within the uncertainty, but beyond the disk, Schwarzschild's light-traces-mass curve decreases much more rapidly. The lower two curves are theoretical rotation curves for the Milky Way Galaxy based on neutral hydrogen observations at Dwingeloo as fitted by a mass model due to Schmidt. These are very uncertain, particularly in the outer regions. The underlying assumption of the models is that the mass converges to a finite value. (With permission of H. van Woerden.)

populated north of the Netherlands. With this instrument an initial map was made of the neutral hydrogen distribution in the northern part of the Milky Way, and the rotation curve of the Galaxy was derived (see Fig. 2.4). But, because we are sitting in the Galaxy, the rotation curve could only be determined with reasonable reliability out to the position of the Sun, about 8 kpc from the center of the Galaxy. Moreover, because of dust obscuration, it was impossible to determine the true light distribution in the Milky Way. So the question of whether or not light traced mass could not be addressed in our own Galaxy: this required accurate measurements of the rotation curves of external galaxies and this, of course, required not only considerable sensitivity but also high spatial resolution. At that time the 21-cm line could be detected and mapped with reasonable resolution only in the very nearest spiral galaxies, which brings us back to M31.

In 1957 Henk van de Hulst, Ernst Raimond, and Hugo van Woerden published the first 21-cm line observations of M31 showing the distribution and motion of the neutral hydrogen. They found that the neutral hydrogen extended considerably further than the bright optical image, and from these observations they were able to observe the clear signature of rotation out to 2.5 degrees or about 35 kpc from the center of M31. The significant aspect here is that the rotation curve could, for the first time, be measured beyond the optical image of the galaxy – something just not possible with optical spectroscopy (by definition).

These observations consisted of a set of line profiles along the major axis of the galaxy, so the problem they confronted was how to convert these line profiles into a rotation curve. They noted that the rotation velocity "is of the order of 200 to 250 km/s throughout this region (beyond one degree from the center of the galaxy) and that it does not strongly decrease with increasing distance from the center. A more precise determination requires a calculation of the model line profiles which takes the radiation from all areas inside the antenna beam into account." In their more precise model they derived a slowly declining rotation curve ($1/r^{0.2}$), but far from a Keplerian decline ($1/\sqrt{r}$), beyond the visible disk. This is shown in Fig. 2.4 (from the paper of van de Hulst *et al.*, 1957) where the 21-cm line rotation curve is compared to that of Schwarzschild calculated from the visible-light distribution with the assumption of constant M/L. We see that within the optical disk ($r < 20$ kpc) the two curves agree certainly to within the errors. But beyond, the curve calculated from the light distribution declines much more rapidly than the 21-cm curve. Unlike Babcock and Mayall, who were constrained to the inner regions by their dependence on optical emission lines, the Dutch radio astronomers had found clear evidence for an increasing mass-to-light ratio in the outer regions, but this was not stated explicitly.

In an accompanying article, Maartin Schmidt, the future discoverer of quasars, calculated a mass model of M32 in order to match this rotation curve. A mass model consists of an assumed distribution of mass for which a force distribution can readily be calculated using Newton's law of gravitation. Schmidt's model consisted of two flattened spheroids to represent the disk and bulge of the galaxy. These spheroids, of course, have finite mass (the alternative was not at all plausible in those days) and this leads inevitably to a declining rotation curve (Schmidt does point out that extrapolation of the mass distribution beyond the observations is not justified). He specifically addressed the issue of whether the light distribution traces the mass distribution and concluded that, within the uncertainties of the measured light and mass distribution, this possibility could not be ruled out. This conclusion was, however, certainly influenced by the assumption of a finite-mass model. In any case for the next 15 years or so, this issue was effectively put to rest.

Three additional developments should be mentioned in this discussion of the 30 years following Zwicky's proposal. In 1950 the Swiss astronomer Rudolph Kurth from Bern, writing in the German language *Zeitschrift fuer Astrophysik*, estimated the mass of the Milky Way system out to a large distance (about 30 kpc) using the observed kinematics of globular star clusters (the line-of-sight velocities determined by Mayall and Kinmann). He found that the mass was about three times larger than that estimated by Schmidt on the basis of his mass model for the Galaxy. In other words, it appeared that the mass of the Milky Way Galaxy increased almost linearly with distance from the center.

In 1959, Franz Kahn and Lodewijk Woltjer, at the time both in Princeton, pointed out that the Milky Way and its nearest large neighbor, M31, are approaching each other with a velocity of 300 km/s. Presumably these two galaxies were originally partaking in the Hubble expansion of the Universe and moving away from each other. So somehow they managed to reverse their separation velocity and are now falling together. Gravity is the obvious mechanism for the reversal, and this would be possible, Kahn and Woltjer estimated, if the mass between the two galaxies were 10 times greater than that in observable stars and gas. They realized that this dynamically necessary mass had to be essentially invisible and suggested that it may be in the form of ionized hydrogen filling the local group of galaxies.

In 1960, Jan Oort, following up on work that he had first done in 1932, considered the distribution and velocity of luminous giant stars perpendicular to the plane of the Milky Way Galaxy. Oort had realized that these observations could be combined to yield an estimate of the gravitational force perpendicular to the plane, and this, in turn, is related to the average mass density in the plane. Oort calculated that this should be 10×10^{-24} gm/cm^3 which would correspond to about six hydrogen atoms per cubic centimeter. The known stars could account for about two atoms per cubic centimeter and the interstellar medium for perhaps about two more. So only about two-thirds of the dynamically present mass in the plane of the galaxy could be accounted for by known forms of matter. In contrast to his earlier analysis (1932), Oort concludes that the discrepancy is real, but, in general, this was not considered to be a serious discrepancy. It was thought that there could be an additional undetected component of the interstellar medium; molecular gas perhaps. Unlike the discrepancy in the Coma cluster, it did not require a dominant invisible component.

2.5 Finzi sums it up

In 1963, 30 years after Zwicky's discovery of the virial discrepancy in the Coma cluster, all of these results were very nicely summarized in a remarkable, but largely forgotten, paper – a true masterpiece of modern astrophysical reasoning – by Arrigo Finzi, then at the University of Rome. Following Kurth, Finzi noted that in the Milky Way, the mass enclosed within a certain distance from the center seems to grow with that distance ($M(r) \propto r$), at least beyond the position of the Sun: three times further away than the Sun, the interior mass was three times larger. He also noted that the rotation curve of M31, as measured in the 21-cm line at Dwingeloo, appears to decline much more slowly beyond the visible disk than the rotation curve predicted by Schwarzschild from the visible-light distribution assuming constant M/L; that in terms of Newtonian dynamics, the outer regions of M31 seem to be increasingly darker. And then finally, on the largest scale, the great clusters of galaxies required mass-to-light ratios in excess of several hundred for virial stability. He, in fact, had put his finger on the ubiquity of the mass discrepancy in astronomical systems – a mass discrepancy that seems to increase with scale.

Finzi's paper had essentially no impact for two reasons: first of all, he was at least 10 years ahead of his time in his understanding of the generality of the problem. Secondly, his solution to this puzzle was unconventional: he proposed a modification of Newton's law in which, beyond the scale of about 1kpc, the force fell like $1/r^{1.5}$ instead of $1/r^2$. He discussed the possibility of dark matter in clusters, but only the classical (baryonic) forms of dark matter (no one could have possibly imagined non-baryonic dark matter at that time). He considered and dismissed four possibilities. First the unseen mass could be in the form of ionized hot gas with a thermal velocity equal to the random velocity of the galaxies (1000 km/s) which would correspond to a temperature of about 10^7 K. But at the density required to provide the unseen mass, the cooling time would be too short (this component, as we shall see, was later discovered but not of sufficient mass to remove the discrepancy). Second, intergalactic stars; but these must have a very different distribution by mass than the stars near the Sun (weighted toward lower-mass stars), and they would not be expected to form in a tenuous intergalactic medium. Third, neutral gas, but this was already ruled out by 21-cm observations (upper limits of 10^{12} $M_\odot$ whereas 10^{15} $M_\odot$ was needed). And finally, solid particles like grains or meteorites, but these are composed primarily of heavy elements which comprise too small a fraction of the mass budget of the Universe to make up 90% of the mass of Coma. So he concluded that the most plausible explanation was a different law of gravity on these scales. Apart from this unconventional suggestion, his was a truly far-sighted recognition of the reality and nature of the discrepancy.

In summary, it can certainly be said that Zwicky, applying Newtonian dynamics, discovered the large discrepancy between the visible and dynamical mass in the Coma cluster of galaxies and proposed that most of the matter in Coma was dark. He first used the term "dark matter" in its modern context. Zwicky realized that the dark matter could be associated with the individual galaxies or could be more smoothly distributed as an intergalactic medium. The early work of Babcock on M31 suggested that individual galaxies exhibited a similar discrepancy with the mass-to-light ratio increasing with distance from the center of the galaxy (although Babcock appeared reluctant to use the term "dark matter"). This was countered by Schwarzschild, who argued that the optically derived rotation curve of M31 was completely consistent with a constant mass-to-light ratio – that galaxies did not become *darker* in the outer regions. However, Schwarzschild did estimate (erroneously) extremely high mass-to-light ratios of elliptical galaxies (> 100); he realized that this could not be made up of stars with a normal mass distribution. By suggesting that the high M/L could be due to dead stars – cool white dwarfs – in a sense, he was attempting to identify the dark matter, although this was not the language that he used. He did point out that with such high M/L values for individual galaxies, the discrepancy in the Coma cluster became less severe, and speculated that it would probably vanish with more accurate observations.

The very first 21-cm line observations of M31, made at Dwingeloo in the Netherlands, permitted a determination of the rotation curve beyond the optical disk, a rotation curve which declined far less rapidly than predicted from the light distribution. But, preconceptions that the mass of a galaxy should be finite, and subsequent model fitting, did not identify this as a problem or a discrepancy. The most important aspect of this work is that within the optical disk, the observed rotation curve is consistent with that calculated by Schwarzschild from the distribution of visible light; the significant discrepancy identified later in galaxies appears beyond the visible disk where the rotation curve, if light traces mass, should be declining but is not observed to be.

It is fair to say that, in the 30 years after Zwicky's discovery, with the exception of a few individuals like Finzi, there was, for astronomers, no sense of a crisis. Gravitationally bound systems from galaxies to clusters could be well described in the context of Newtonian dynamics by the observable matter, perhaps augmented by familiar forms of matter such as dead stars. In fact, the sense of crisis emerged not from observations but from theoretical considerations.

3

The stability of disk galaxies: the dark-halo solution

3.1 Building disk galaxies: too hot to be real

In the early 1960s, computing power, measured either in terms of calculations per unit time or rapid access memory capacity, appeared to undergo an enormous, almost discontinuous, leap forward. This development resulted primarily from the replacement of vacuum tubes by transistors, and now, viewed on the timescale of a century, we know that it is only one segment of an exponential curve describing the time evolution of computing power – a phenomenon encapsulated in the famous Moore's law: by any means of measuring it, computing power doubles every two years. With respect to theoretical astrophysics, this meant that by 1960 it had become practical to apply electronic computing machines in the numerical solution of complex problems such as solving for the structure and evolution of stars or the transfer of radiation at various wavelengths through stellar atmospheres.

By mid-decade several innovative astrophysicists and dynamicists were considering the computer solution of the Newtonian N-body problem where N was considerably larger than a few – in fact, on the order of 100 000. The problem is straightforward: set up a system of particles each with a prescribed mass, calculate the Newtonian gravitational field generated by these particles, and then let them move under the influence of this force field for a short interval of time. Of course, after this interval, because the particles have rearranged themselves, the force has to be recalculated before the particles are moved further. It sounds simple, but to do this accurately for a large number of particles requires a computer with fast processing, a great deal of rapid access memory and very clever programmers.

Among the pioneers in attacking this problem were Richard Miller at the University of Chicago, Kevin Prendergast at Columbia University and Frank Hohl of the NASA Langely Research Center. They were the first to consider the problem of two-dimensional N-body systems, the kind of system which might be relevant to flat-disk galaxies. All of the particles were distributed in a plane but, of course, the gravitational force between the particles was still three dimensional. The entire system was rotating about its center of mass; the centrifugal force balanced the force of gravity and prevented the entire collection of particles from collapsing into one lump at the center. This rotational support appears to be the case in real spiral galaxies like the Milky Way. As we have seen, these are massive stellar systems which are rotating with velocities of typically 100 to 300 km/s. Miller and Prendergast (1968), and independently Hohl (along with R.W. Hockney in 1969), wanted to simulate such systems. One of their goals was to see if the glorious spiral structure observed in disk galaxies would develop naturally as waves in the density and gravity field of numerical disks.

They wrote their programs on decks of cards, one command per card, read them into their enormous university computers (which in those days occupied large rooms filled with tape drives, card readers, printers, cables, air conditioners, and the machines themselves with hundreds of flashing lights) and observed their numerical galaxies. Of course, they had an advantage over real astronomers who observe galaxies in the sky. These were actual numerical experiments and the theoretical astronomers became experimentalists. They could change the initial properties of the system or disturb it in various ways to see how it would respond.

Fig. 3.1 is an example of what they found. This is from the work of Frank Hohl (1971), but the results are very similar to those of the somewhat earlier calculations by Miller and Prendergast. The experiments began with a disk of particles supported in equilibrium almost entirely by rotation. Equilibrium means that for every particle, the gravitational force pulling it in toward the center is almost balanced by the centrifugal force pushing it outward. Therefore, the global structure of the system should not change: the particles should simply continue on their nearly circular paths about the center of the simulated galaxy.

But, to the surprise of everyone, this was not what happened. Rapidly, on the timescale of a few rotation periods, the computer galaxy changed its form and evolved into a more elongated shape – more like the barred spiral galaxy seen in Fig. 3.2. Almost one-half of spiral galaxies show evidence for such bar structures which suggests that bars are long-lived configurations. However, the bars created in these early numerical experiments kept on evolving. The bar seemed to dissolve but the orbits of the individual particles remained very elongated; that is to say, the paths of particles were more similar to elongated ellipses than to circles, but the long axes of the ellipses were at random angles to each other.

Such a rapid evolution of a system in equilibrium is known as an instability. So, the numerical experiments seemed to be telling us that a cold rotationally supported disk is globally unstable to the formation of an elongated, or "non-axisymmetric" system. There appear to be non-axisymmetric instabilities that cause the system to evolve from a disk shape to a bar shape. The further evolution then suggests

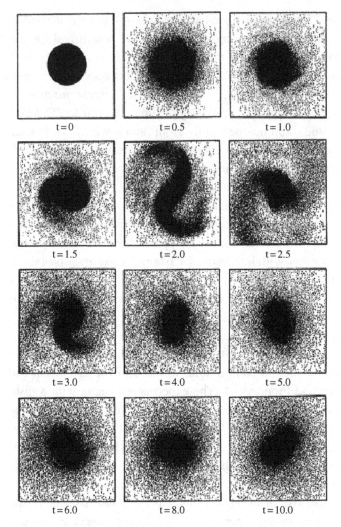

Fig. 3.1. The time evolution of a cold disk of stars. The indicated timescale is in units of the average rotational period of the system. The initially axisymmetric system is seen to rapidly evolve into a bar shape which then appears to dissolve somewhat. From calculations by Frank Hohl (1971).

a return to a rounder or more axisymmetric disk form, one in which the individual particles are no longer on circular orbits but traverse highly elongated paths with large excursions in radius. That is to say, the system evolves from being rotationally supported to being pressure supported, like the gas that inflates a balloon.

This was quite a surprising result, because a galaxy like the Milky Way does not look that way. The orbits of the stars near the Sun appear to be very nearly circular; they are going around the center of the Galaxy with a circular velocity in excess of 200 km/s. On top of that circular velocity there is also a random velocity – the

Fig. 3.2. The barred spiral galaxy NGC 1300. A large fraction of disk galaxies show such features suggesting that they are long-lived stable features of disk galaxies, unlike the bars appearing in the original numerical galaxies (Hubble Space Telescope image, courtesy of NASA).

velocity vectors point in all directions – but that is only 30 or 40 km/s. So our Galaxy seems to be supported almost entirely by systematic rotation, at least in the neighborhood of the Sun. How is it that a real galaxy can be cold and rotationally supported, but a numerical galaxy supported by rotation is unstable and rapidly evolves toward a pressure-supported system?

These early experimental astrophysicists were very puzzled by these results, but because the calculations really were numerical experiments they could try various "fixes" to make the simulated galaxies look more like real galaxies. One of these fixes, tried by Miller and Prendergast, then joined by their student Bill Quirk, was to artificially cool the heated system of particles: at every time step, they reduced the amount of energy in the random motions of the particles by an arbitrary factor. This is not quite as ad hoc as it might seem because galaxies not only consist of stars but also of gas. The gas is apparently clumpy; that is, the gas is not uniform but distributed in clouds of all sizes. Unlike the stars, the gas clouds can collide and lose energy. So such artificial cooling, in fact, simulates the energy loss, the "dissipation" of real gas clouds in a galaxy.

One of the most interesting consequences of this cooling was the very clear appearance of spiral structures, seen in Fig. 3.3 from the calculations of Miller, Prendergast and Quirk (1970). Only a fraction of the particles are allowed to cool, to simulate clouds. The panel on the left shows the distribution of these "gas" particles, and we can see a clear spiral structure. The panel on the right, however, shows the distribution of "stars"; those particles with no artificial cooling which move only according to Newton's laws. The appearance of this system of uncooled

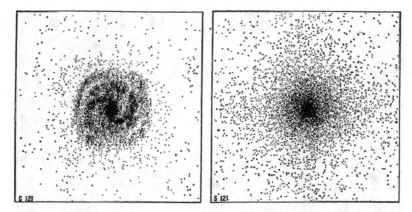

Fig. 3.3. The effects of artificially cooling a fraction of the particles in the calculations of Miller, Prendergast and Quirk (1970). The left panel shows the distribution of cooled particles (the gas) and the right panel shows the distribution of uncooled particles (the stars).

particles is much more axisymmetric and with much more random motion; much hotter than the system of gas particles. Although it is not apparent on the figure, the system of stars also exhibits a weak spiral structure coincident with that of the gas particles.

These spiral arms are really density waves – the particles become a bit more crowded as they move through the pattern. This had been proposed several years earlier, by C. C. Lin and Frank Shu of MIT, as a solution to the "winding up" problem of spiral structure. But, the cooling of the "gas" component did not solve the overall problem of disk heating. As soon as the cooling was turned off, the instabilities grew again with the system of collisionless particles developing a much larger random motion than observed in the Galaxy.

The results of the numerical experiments on disk galaxies were summed up neatly by Frank Hohl (1971): "Attempts to slowly cool axisymmetric stable (i.e., hot) disks indicated that.... further cooling (beyond some minimum velocity dispersion) would only cause collective instabilities which would heat up the disk as fast as it was being cooled." The tension between observations and calculations was very sharp indeed. Somehow real galaxies found a way of maintaining stable rotationally supported disks. Something was missing in the computer simulations.

3.2 Dark halos to the rescue

Jerry Ostriker was at that time a young assistant professor at Princeton University. His thesis adviser at the University of Chicago had been the famous Subrahmanyan Chandrasekhar, the discoverer of the upper limit to the mass of white dwarf stars. Ostriker had worked on the effects of rotation on white dwarfs and the possible impact of rotation on the Chandrasekhar limit. Could the mass limit be extended to higher masses and so avoid the black-hole trap of gravitational collapse? Because of this work he was very familiar with the physics and the classical theory of rotating fluid spheroids.

But his interests were far more wide-ranging than rotating stars. He had also worked in the field of galaxy dynamics and was very aware of the contradiction between the apparent instability of the rotationally supported computer-simulated disk galaxies and the obvious stability of the rotationally supported real Milky Way Galaxy. And he had a completely new idea about how this conflict might be resolved.

Consider the discussion of the virial theorem given in the Appendix (Section A4). For a system in equilibrium it must be the case that

$$2T + U = 0 \tag{3.1}$$

where T is the total kinetic energy of the system – the energy in the actual motion of the particles – and U is the total (negative) gravitational potential energy – the energy required to disperse the system to infinite distance. This is true of any system in equilibrium – a star, a galaxy, a cluster of galaxies. We may divide the kinetic energy into two parts: the kinetic energy in directed rotational motion T_{rot}, and the kinetic energy in random motions, T_{ran}. This final bit would be the "heat" energy of the system. Then, with some rearrangement, the virial theorem becomes

$$T_{rot}/(-U) + T_{ran}/(-U) = 1/2. \tag{3.2}$$

Defining $t = T_{rot}/(-U)$ and $r = T_{ran}/(-U)$, the virial theorem may be written

$$t + r = 1/2. \tag{3.3}$$

So if $t = 1/2$ ($r = 0$) the system is completely supported against gravity by rotation, but if $r = 1/2$ ($t = 0$) the system is completely supported by random motion. Thus t is, in a sense, a measure of the temperature of the system; although in this case high t means "cold" and low t means "hot".

Ostriker knew from his study of classical rotating spheroids that whenever t is greater than about 0.14; i.e., when more than 28% of the kinetic energy of the system is in rotational motion, the spheroid would be unstable, and this instability leads to the formation of an elongated (prolate) object. The instability appears at this point because it is energetically favorable for the system: by increasing its moment of inertia, the system can decrease its rotational energy while conserving its angular momentum. Ostriker also realized that the numerical galaxy disks do in fact initially violate this stability condition, so if the same criterion applies to stellar systems as well as to fluid spheroids, rotationally supported disks really should be unstable, as the N-body experiments suggested. But then so should the Milky Way! As mentioned above, the stars of the Milky Way, at least near the

Sun, have a random velocity of 40 km/s in addition to a rotational velocity of 200 km/s. Since the kinetic energy is proportional to the velocity squared then, with a little bit of arithmetic, $t \approx 0.49$ for the Galaxy, far in excess of the stability limit. The Galaxy should be violently unstable. How can this problem be solved?

Suppose, reasoned Ostriker, that the Milky Way really is a hot system. Suppose that there is an additional component to the Galaxy, a spheroidal hot component, extending far above the plane of the Galaxy, which contributes at least 50% of the mass inside the position of the Sun. Then this spheroidal system would add to the gravitational potential energy, but add nothing to the rotational energy; t would be decreased and perhaps stability restored. The disk stars would be moving in near circular orbits in the gravitational field of the spheroidal component much as the planets move in the gravitational field of the Sun. There is of course no problem with the stability of the Solar System; at least, fortunately, not on the timescale of many, many planetary-orbit periods.

A spheroidal component of the Galaxy is, in fact, observed – a luminous system of globular clusters and a few high-velocity stars with a roughly spherical shape extending out to tens kpc from the center of the Galaxy. But if we add up the mass of this stellar component, we find that it is only a few percent of the mass of the disk; it falls far short of what would be necessary for stability of the disk. So where is the needed spheroidal component? Here Ostriker made a giant logical leap. Suppose the spheroidal component indeed exists but has a much higher mass-to-light ratio than the stars in the disk or in the luminous spheroid. Suppose that the Galaxy is embedded in a massive "dark halo".

To consider the effect of a rigid spheroidal halo on the stability of the disk, Ostriker joined with James Peebles (1973), a cosmologist at Princeton who would later make his own very major contribution to the cosmological dark matter problem. Peebles had an N-body program, and he and Ostriker carried out several numerical experiments on disks consisting of a few hundred particles. This was a much smaller number of particles than had been considered by Miller, Prendergast and Hohl in their experiments, but they were looking specifically at the problem of disk stability and the effects of adding a rigid spherical component to the gravitational field.

Ostriker and Peebles found that, indeed, pure disks supported by rotation rapidly changed their form: the initially axisymmetric disks rapidly evolved into an elongated shape and then dissolved into a hot pressure-supported disk. In particular, the fraction of energy in rotation rapidly decreased from nearly 0.4 to about 0.14 as had been supposed by Ostriker. But when they added a rigid spherical halo, represented by an additional component of the gravitational force that did not correspond to the actual particles, they found that the disk maintained its rotational support and its

Fig. 3.4. The Ostriker–Peebles–Yahil (1974) view of spiral galaxies. The galaxy disk is embedded in a more extensive dark pressure-supported system – a dark halo with a total mass larger than that of the visible object.

axisymmetric shape provided that the halo mass was at least equal to the disk mass. Basically t for the combined halo and disk had to be near to or less than 0.14 for stability of the disk.

This is shown in Fig. 3.5, reproduced from the work of Ostriker and Peebles; this is a plot of t against time in units of the orbit period for the outermost particles. We see that when there is no halo, t decreases from an initial value near 0.4 down to about 0.14. But when the halo mass is at least twice the disk mass, t is initially near 0.15 and remains there. A halo does indeed provide the necessary stability.

So it appears that Newtonian rotationally supported disks require a spheroidal halo for stability. Because no such halo is seen, the halo must be dark, in the sense that the mass-to-light ratio of whatever makes up the halo should be much higher than that of a "normal" stellar population. After all, there is no reason why every gram of matter in the Universe must produce about one erg per second as the Sun does. Ostriker and Peebles speculated that the halo could be made up of very low-mass stars, or white dwarfs – objects that do have a very high mass-to-light ratio. They suggested systematic searches for such low-luminosity objects – objects that would emit most of their radiation at infrared wavelengths and have a very high velocity with respect to the Sun.

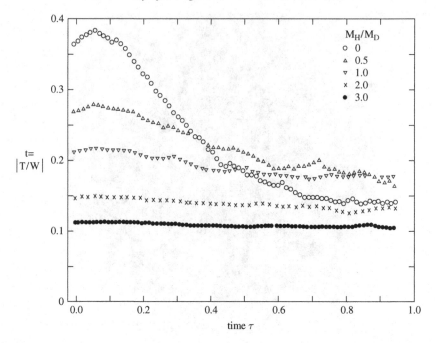

Fig. 3.5. The effect of the halo on the evolution of the model galaxy. The fractional energy in rotation is plotted against time for various values of the halo-to-disk mass ratio. With no halo, the rotationally supported disk ($t \approx 0.4$) rapidly heats up to a level where the disk is primarily pressure supported ($t \approx 0.14$). When a halo with a mass comparable to the disk mass is added (in the form of a rigid gravitational field) the disk barely warms (from Ostriker and Peebles, 1973).

Ostriker and Peebles published their results in 1973, and the suggestion that every spiral disk is embedded in a dark halo was, at the time, quite controversial. Of course, to provide stability to the Milky Way, this dark halo does not really need to extend much beyond the position of the Sun, or be much more massive than the disk. For the purposes of stabilizing galaxy disks, perhaps the amount of matter in the Universe need only be doubled.

3.3 Larger implications

Two years later, Ostriker and Peebles, joined by Amos Yahil, looked at the problem of dark mass in a more general sense, in the sense explored by Finzi a decade earlier. Dark halos were only one aspect of the dark matter problem – a problem apparently existing in all astronomical systems on the scale of galaxies and larger.

They first returned to the old argument of Kahn and Woltjer (1959): the Milky Way and its neighbor, M31, are approaching each other much too rapidly if their mutual gravitational attraction were only due to the visible matter in each galaxy.

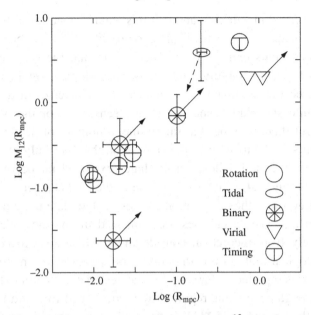

Fig. 3.6. The mass of local spiral galaxies in units of 10^{12} M$_\odot$ (i.e., a unit roughly equal to 10 times the visible mass of the Milky Way), as a function of distance from their centers (in units of Mpc). The mass is estimated by various dynamical arguments. Ostriker, Peebles and Yahil argue, on the basis of this plot, that the total mass of such systems continues to increase linearly with scale well beyond that which is necessary for stability.

Kahn and Woltjer tried to explain this discrepancy as being due to warm gas between the two galaxies, but this observation fits perfectly well with the idea that each system contained far more invisible than visible mass. Ostriker, Peebles and Yahil (1974) then considered a number of additional observations. The dynamical mass on a range of scales is probed by satellite galaxies of the Milky Way, by galaxies apparently orbiting about one another – double galaxies – by small groups of galaxies similar to the "local group". Their plot of mass vs. scale is reproduced in Fig. 3.6. It appears that on larger and larger scales, the enclosed mass increases linearly with scale; more and more dark mass is required to bind larger systems together. This would be the expected result if the dark halo required for disk stability extends far beyond the inner regions, if the interior mass in such a halo (and hence the interior mass-to-light ratio) kept on increasing with distance from the center of the galaxy. The mass-to-light ratio of spiral galaxies out to several hundred kpc would then be comparable to the average mass-to-light ratio of elliptical galaxies in the Coma cluster (estimated long before by Zwicky); the large mass-to-light ratio in clusters is thus explained by the dark halos of galaxies. They pointed out that the quantity of mass in the Universe could be 10 times, or 100 times, that directly observed in galaxies, and that this would be sufficient to increase the density of the Universe to its critical value, to provide $\Omega = 1$ (see Section A5).

This is an enormous extrapolation. The idea is that the same dark material required for disk galaxy stability might be of cosmological significance – that it might, in fact, be the dominant constituent of the Universe. In the mid-1970s, this seemed quite speculative and not in any sense part of the world view of astronomers. But then, at about the same time, a number of actual observations appeared which seemed to support the suggestion that, not only were the visible galaxies immersed in dark halos, but the halos extended far beyond the visible disks of galaxies (Fig. 3.4). That is the subject of the next chapter.

But before I go on to the observations, let us contemplate these developments. The suggestion that visible galaxies are immersed in enormous dark halos has originated largely from theoretical considerations. Of course, there is a history of phenomenological hints in this direction – observations of "missing mass" in clusters of galaxies reported by Zwicky; an undetected contributor to the force perpendicular to the galactic plane reported by Oort; a need for extra mass in order to understand the approach of M31 to the Milky Way considered by Kahn and Woltjer. Moreover, as I discuss in the next chapter, observations of the extended rotation curves of spiral galaxies were just beginning to appear in the literature – observations which implied that the mass-to-light ratio was certainly increasing in the outer regions.

But the radical suggestion that galaxies are mostly dark and that the dark material was distributed in a *spheroidal* halo grew out of the tension between observations of the Milky Way stellar velocity field, strongly implying rotational support of the galaxy disk, and the initial N-body simulations which indicated instability of rotationally supported systems. It took someone with a flexible and creative intelligence to make this connection between instability of Newtonian rotationally supported systems and the necessity of a stabilizing pressure-supported spheroidal component to account for the existence of apparently rotationally-supported galaxy disks. Of course, because the hot spheroidal system is not seen, it must be dark.

Subsequent research has shown that the global criterion for stability suggested by Ostriker and Peebles, $t < 0.14$, is an oversimplification and that the Ostriker–Peebles solution to the stability problem (i.e., adding a halo which contributes significantly to the mass within the inner regions of the disk) is not unique. Analysis of N-body experiments by Lia Athanassoula and Jerry Sellwood (1986) demonstrates that high random motion in the central parts of a galaxy also acts to stabilize a disk that may be largely supported by rotation in the outer regions. In this way the halo mass within the position of the Sun required for stability of the Milky Way can be substantially reduced. Nonetheless, an extended halo does seem to be necessary for suppression of all instabilities of rotating Newtonian disks, and, regardless of the exact criterion, Ostriker and Peebles were the first to appreciate the problem of the stability of cold rotating disks and grasp its implications. The solution which they suggested, at least in its extrapolated form with halos extending far beyond the visible galaxy, has, in fact, turned out to be the one subsequently supported by observations. Their proposal has led to a view of spiral galaxies that is entirely different from that which preceded their work: the visible disk is only a small component of a vast dark system extending far beyond the disk.

An important caveat, and one that I have tried to emphasize by frequent use of the word "Newtonian", is that the stability requirement applies to disks which obey Newton's laws of gravity and dynamics. But certainly, a credible alternative to dark halos in the context of Newtonian gravity and dynamics, must also address this issue of the stability of rotating self-gravitating disks. I will return to this point in Chapter 10.

4

Direct evidence: extended rotation curves
of spiral galaxies

4.1 Radio telescopes: single-dish and interferometers

By 1970 radio astronomy had emerged as a major tool for exploring galactic and extragalactic phenomena. The telescope antennae had grown in size and precision of surface from the early primitive World War II radar dishes. Greater size meant greater resolution and sensitivity; radio sources, including galaxies, could be mapped in finer detail and at larger distances. At the same time, the technology of radio receivers was undergoing rapid development; the intrinsic electronic noise of receivers, the background static, was (and still is) being continuously reduced so that any particular dish could detect fainter signals in a shorter observing time.

Notable among the very large steerable single-dish telescopes that had come on line at this point were the 250-foot telescope at Jodrell Bank near Manchester, UK (operated by the University of Manchester), the 300-foot telescope at Green Bank, West Virginia (the National Radio Astronomy Observatory or NRAO), and, by 1972, the 100-m radio telescope in Effelsberg, Germany (operated by the Max Planck Institute for Radio Astronomy, MPIfR, in Bonn, see Fig. 4.1).

The construction of larger and larger single-dish radio telescopes has engineering limitations and, fortunately, is not the only means of increasing resolution. Radio interferometers, the use of an array of antennae covering a much larger area, could, by combining signals, effectively act as one telescope with an enormously increased aperture. This technique was pioneered at Cambridge (UK) by Martin Ryle, and similar radio interferometers were developed at Green Bank by NRAO, at Owens Valley (California) by the California Institute of Technology, and at Westerbork WSRT (the Netherlands) by the Netherlands Foundation for Radio Astronomy (Fig. 4.2).

What kind of signals do radio telescopes detect from galaxies? There are basically two kinds of emission: continuum – radiation emitted over a wide range of wavelengths – and spectral line – radiation emitted at a single discrete wavelength.

Fig. 4.1. The 100-m radio telescope of the Max Planck Institute for Radio Astronomy in Bonn.

Fig. 4.2. The Westerbork WSRT synthesis radio telescope of the Netherlands Foundation for Radio Astronomy.

Both kinds of radiation are detected from galaxies at radio frequencies. The continuum radiation has two sources: thermal radiation, or Bremsstrahlung, which is emitted by free electrons in a hot ionized gas; and synchrotron, which is produced by relativistic electrons, cosmic rays, spiraling in interstellar magnetic fields.

The line radiation emerges from discrete transitions in atoms (or molecules), and in radio astronomy, the most important line with respect to observing the gas kinematics in galaxies is certainly that of neutral hydrogen at a wavelength of 21 cm (Chapter 2). Although it is somewhat of a simplification, we may consider a neutral hydrogen atom as an electron orbiting around a proton. Both kinds of particles have the property of "spin" which is quantized – up or down. Because of the magnetic interaction between the particles, the energy of the configuration is slightly higher when the spin of the electron is aligned with that of the proton than when anti-aligned. The transition from the higher energy configuration to the lower gives rise to the line emission at 21 cm. Because hydrogen is the most abundant element in the Universe and because neutral hydrogen is a major component of the interstellar medium in galaxies, the 21-cm line is the ideal probe of the distribution and motion of gas in spiral galaxies. The motion of the gas toward or away from the observer can be detected by the well-known Doppler shift – the line is shifted to shorter wavelengths (toward the "blue") if the gas is moving toward us and to longer wavelengths (toward the "red") if the gas is moving away (see Section A1).

The resolution of a radio telescope in minutes of arc, the "beam size", depends upon the wavelength of the radiation (in cm) and the size of the dish – the aperture – (in m) as

$$\theta \text{ (arc min)} \approx 34 \lambda \text{ (cm)}/D \text{ (m)}. \tag{4.1}$$

At a wavelength of 21 cm even the largest single-dish radio telescopes, such as the 100-m Effelsberg telescope, generally do not have the resolution ($\approx$ 8 minutes of arc) to map a distant galaxy in detail; except for relatively nearby galaxies like the great spiral in Andromeda (M31), the entire galaxy is usually included in the "beam" of the radio telescope, and so, the telescope detects all of the neutral hydrogen in the galaxy emitting over a range of velocities. This gives rise to a characteristic shape or "profile" for the 21-cm line emission detected by a single-dish telescope observation of a distant rotating spiral galaxy (line profile means the intensity of radiation as a function of frequency or line-of-sight velocity). The formation of such a line profile (Fig. 4.4) is illustrated in Fig. 4.3.

Imagine a galaxy with an inclination of 30 degrees to the plane of the sky. The intrinsic shape of a disk galaxy is nearly circular, so the circular disk projected onto the sky will appear as an ellipse. Fig. 4.3 (upper panel) shows the rotation curve of this hypothetical galaxy. The gas in the galaxy is rotating about the center with

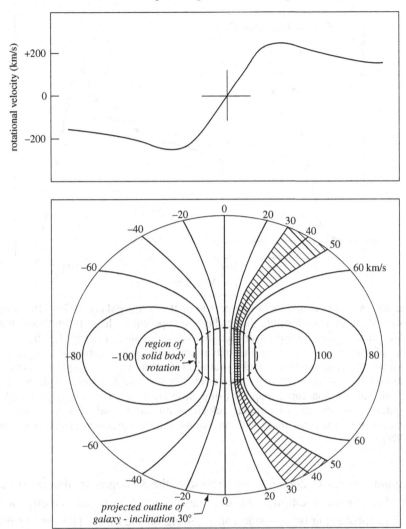

Fig. 4.3. The upper panel is a hypothetical galaxy rotation curve. The lower panel illustrates the resulting two-dimensional velocity field for a galaxy disk inclined by 30 degrees to the plane of the sky. The curves are lines of constant radial velocity (the component of the rotation velocity along the line-of-sight to the observer). The curves are actually blended together because of random motion of gas in the galaxy (turbulence) and the finite velocity resolution of the telescope receiver; the shaded region shows the effect of such blending. Except for the closest galaxies, a single-dish telescope "sees" the entire galaxy in its beam, so the entire range of velocities is present within the 21-cm line. The resulting line profile (the shape of the line as a function of frequency or radial velocity) is similar to that shown in Fig. 4.4 – the characteristic "double-horn" profile. An interferometer has a higher spatial resolution and, for many galaxies, produces such a map of the entire two-dimensional velocity field. From Roberts (1975b).

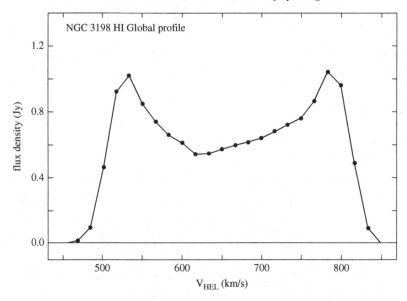

Fig. 4.4. A single-dish 21-cm line profile of the spiral galaxy, NGC 3198. For more distant galaxies a single-dish radio telescope with a large beam "sees" the entire galaxy – the emission over all radial velocities shown in Fig. 4.3. Because most of the emission is near the maximum (projected) rotation velocity, this produces the characteristic "double-horn" 21-cm line profile of spiral galaxies. Although we can estimate the rotation velocity from such a single-dish profile, the detailed rotation curve mapped using an interferometer is clearly preferable in determining the run of circular velocity with radius and, hence, the radial distribution of force (or mass with an assumed law of gravity). From Begeman (1989).

a maximum velocity of 200 km/s but because of the 30-degree inclination the component of maximum rotation along the line-of-sight (the radial velocity) will be 100 km/s, approaching on one side and receding on the other. The lower panel of this figure is a contour map of the radial velocity over the two-dimensional image of the inclined spiral galaxy, i.e., the curves are lines of constant radial velocity. Along the major axis of this ellipse the radial velocities are just the projected rotation velocity. On the minor axis, rotation is perpendicular to the observer's line-of-sight so the radial velocity here is zero (these are the radial velocities left over after subtracting the radial velocity of the entire galaxy, i.e., the "systemic" velocity).

Now imagine that this hypothetical galaxy is observed using a single-dish radio telescope and that this entire object lies within the beam of the telescope. Therefore, the single dish will see one broadened line profile due to the 21-cm line emission over the entire galaxy; in this case the total velocity spread or line width will be about 200 km/s. The emission is particularly concentrated near the projected

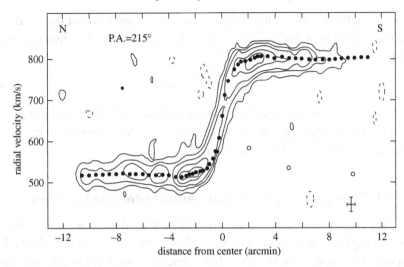

Fig. 4.5. A vertical line (at a given distance from the center) is actually the 21-cm line profile at that point, shown here as a contour map. A stack of such profiles along the major axis of the galaxy image provides a picture of the 21-cm line emission as a function of distance from the center, from which we can derive the rotation curve. This is from 21-cm line observations made at Westerbork of the spiral galaxy, NGC 3198. The fitted rotation curve is shown by the points. From Begeman (1989).

maximum-rotation velocity, toward and away from us, and this gives rise to the global line profile shown in Fig. 4.4 for an actual spiral galaxy, NGC 3198. The 21-cm line neutral hydrogen over a spiral galaxy has a characteristic "double-horned" appearance, and the line width is about twice the projected maximum rotational velocity (about 150 km/s in this case).

An interferometer, because of its much larger effective aperture (thousands of meters), has higher angular resolution than a single dish and for many galaxies can resolve the entire two-dimensional radial-velocity field. By looking at the 21-cm line profiles along the major axis, where the line-of-sight velocities are maximum, the rotation curve can be directly observed if the motion is in fact circular and planar. This sequence of profiles for a real galaxy, again NGC 3198, is shown in Fig. 4.5 as a contour plot. The contours, representing equal levels of 21-cm line intensity, are shown as a function of radial velocity (vertical axis) over a range of angular distance from the center (horizontal axis) along the major axis of the projected disk. The rotation curve may be derived from such a plot, and in this case, is plotted by the sequence of points on top of the contour map (observed at Westerbork WSRT by K. G. Begeman).

Global profiles determined by single-dish radio telescopes do provide a reasonable estimate of the magnitude of the rotation velocity of a spiral galaxy. If i is the angle between the observer's line-of-sight and the rotation axis of the galaxy, then the maximum rotation velocity in the galaxy V_{rot} is related to the observed velocity width of the line ΔV as

$$V_{rot} = \frac{1}{2}\Delta V/\sin(i). \tag{4.2}$$

However, to derive the detailed distribution of force, and, by Newtonian gravity, the mass distribution, it is necessary, in all but the nearest galaxies, to measure the rotation curve with a radio interferometer.

4.2 Early results of single-dish neutral hydrogen observations

In the 1970s two astronomers, Brent Tully, an optical astronomer then at the University of Maryland, and Rick Fisher, a radio astronomer at the National Radio Astronomy Observatory, used single-dish telescopes such as the 300-foot dish at Green Bank to measure the global line profiles of 10 nearby spiral galaxies with known distance (see Section A2). Given the width of the global line profiles and the apparent inclinations of the disks, they could then estimate the characteristic rotation velocity. Knowing the distance and the apparent brightness (magnitude) of these galaxies in the B band (see Section A1) Tully and Fisher (1977) could calculate the luminosity in blue light. So the final result of these observations and analyses was a table of luminosities and rotation velocities for these several spiral galaxies.

Remarkably, they found a tight correlation between the luminosity and the rotation velocity, a correlation that has since become known as the Tully–Fisher relation. Once calibrated on nearby galaxies, this relation becomes an important distance indicator and has played a major role in the determination of the Hubble constant. The original Tully–Fisher relation is reproduced in Fig. 4.6; there appears to be a quite small scatter about a straight line on this log–log plot. This implies a power-law correlation of the form $L \propto V^{\alpha}$ – the tightest empirical relationship in extragalactic astronomy. The Tully–Fisher relation is actually very relevant to the problem of dark matter in galaxies. This is because the luminosity is proportional to the mass of the stellar disk, but the rotation velocity is apparently set by the dark matter beyond the disk. Any theory of galaxy formation or evolution in the context of dark matter must account for this near-perfect correlation between visible and dark matter properties, but this is a topic for later chapters.

In spite of relatively low resolution, the early single-dish 21-cm line observations revealed previously unknown aspects of the gas distribution and motion in galaxies; for example, the neutral hydrogen most often extends well beyond the visible image of the galaxy; the gas disk is larger than the luminous stellar disk. This observational result, based upon a decade of work, was described by Morton

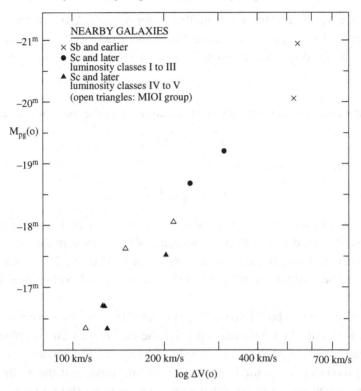

Fig. 4.6. The original Tully–Fisher relation; the correlation between luminosity and width of the 21-cm line, $L \propto V^{\alpha}$, is evident in this log–log plot. From Tully and Fisher (1977).

Roberts (NRAO) in 1975. Roberts (1975a) pointed out that the hydrogen typically extends to 1.5 times the optical radius. This provides the opportunity of measuring the rotation curve of the neutral gas well beyond the optical image of the galaxy – the *extended* rotation curve. What, then, are the expectations for such an extended rotation curve if visible light traces mass?

Although the neutral hydrogen is easily detected by radio telescopes, it actually makes up a very small fraction of the mass of most luminous spiral galaxies. Because the neutral gas does not contribute significantly to the mass budget of galaxies like M31, and because it is extended well beyond the optical image of the galaxy (and presumably most of the mass), then its rotation velocity should be a *tracer* of the gravitational force law beyond the galaxy, just as planetary motion in the Solar System traces the force law beyond the Sun.

Assuming that the Newtonian gravitational force, g_n, provides the acceleration due to circular motion (the centripetal acceleration, see Section A4) we have

$$g_n = V_{rot}^2 / r. \qquad (4.3)$$

So measuring V_{rot} at a distance r from the center of the galaxy, we can determine the gravitational force g_n. If the measurements are made well beyond most of the mass M of the galaxy, then, according to Newton's law, the gravitational force is

$$g_n = GM/r^2. \tag{4.4}$$

Combining these two formulae provides an estimate of the mass of the galaxy

$$M = V_{rot}^2 r / G. \tag{4.5}$$

We can also turn this formula around:

$$V_{rot} = \sqrt{GM/r} \tag{4.6}$$

or, in other words, the rotation curve should fall like $1/\sqrt{r}$ in the outer parts of the galaxy, just like the decline of orbital velocity of the planets in the Solar System. So the expectation is that the rotation curve measured in the 21-cm line of neutral hydrogen beyond the visible edge of the galaxy should decline in a Keplerian fashion.

This expectation was built into the early models of spiral galaxy rotation curves – models such as that which Schmidt applied to the very first 21-cm line observations of M31 made at Dwingeloo (as described in Chapter 2). Such models were based upon the preconception, entirely reasonable at the time, that the visible galaxy defined the extent of the mass distribution, and so there should be a Keplerian decline in the rotation curve. But this is not what was observed.

In the early 1970s Morton Roberts joined by Robert Whitehurst of the University of Alabama determined the extended rotation curve of M31 from 21-cm line observations made at the 300-foot antenna. In work not published until 1975, they noted that the neutral hydrogen in M31 extends far beyond the point where starlight can be detected – in fact, almost twice as far. But, surprisingly, the rotational velocity measured in this thin hydrogen gas beyond the bright galaxy appears to be equal to that in the inner regions and not declining at all – certainly not in a Keplerian fashion ($\propto 1/\sqrt{r}$). The rotation curve is flat, completely contrary to expectations (this absence of a Keplerian decline beyond the visible disk had already been suggested by the early 21-cm line observations of the Dutch radio astronomers as discussed in Chapter 2).

What are the implications of such a flat rotation curve? Looking back again at eq. 4.5 above, we can now replace the constant M by $M(r)$; that is, we are supposing that the mass enclosed within r is not constant but a function of distance from the center. The rotation velocity is constant, so eq. 4.5 tells us that $M(r) \propto r$. That is to say, the enclosed mass goes on increasing as distance from the galaxy center. However, the total enclosed luminosity, $L(r)$, does not go on increasing as distance; it converges to a fixed value, L. In 1970, Kenneth Freeman of Mt. Stromlo

Observatory in Australia published a quantitative analysis of the light distribution in spiral galaxy disks and pointed out that, in general, the surface brightness, I, falls exponentially with distance from the center:

$$I = I_0 e^{-r/h} \tag{4.7}$$

where I_0 is the central surface brightness and h is the characteristic scale size of a galaxy (Freeman also noticed that, unlike the scale length h, the central brightness I_0 seemed not to vary much from galaxy to galaxy). This means that the entire luminosity of the galaxy is contained within radii of three or four scale lengths. Freeman calculated the Newtonian rotation curve expected for an exponential disk and demonstrated that it should decline in a Keplerian manner beyond about three disk scale lengths (Freeman in 1970 had already suggested, on the basis of the observed non-declining rotation curves of two galaxies, that there was evidence for an increasing M/L in the outer regions).

So the total visible starlight and, presumably, the total density of stellar matter falls off very sharply with distance from the center of the galaxy but the mass density does not. If the enclosed mass increases as radius, then it is easy to demonstrate that the mass "surface density" (solar masses per square pc) falls as $1/r$. But the light is decreasing exponentially. This gives rise to a dramatic increase in mass-to-light ratio in the outer regions of the galaxy as is illustrated in Fig. 4.7 for a hypothetical galaxy disk with a flat rotation curve. Here the solid curve is a plot of the exponential decrease of surface brightness (I) in solar luminosities per square parsec. The dotted curve shows the decline of surface density ($M_\odot/pc^2$) and the dashed line is the resulting mass-to-light ratio in solar units. It would appear that

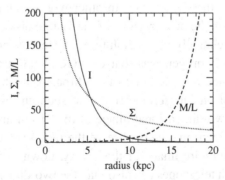

Fig. 4.7. The surface brightness (solid curve, $L_\odot/pc^2$), mass surface density (dotted curve, $M_\odot/pc^2$) and mass-to-light ratio (dashed curve, solar units) as a function of radius for a hypothetical disk with an exponential light distribution and a flat rotation curve. It has been assumed here that the exponential scale length (h) is 2 kpc.

spiral galaxies were becoming darker and darker in the outer regions. Could this really be true?

4.3 Early observations of spiral galaxies with radio interferometers

The largest single-dish radio telescopes, such as those at Green Bank or Effelsberg, have a resolution of the order of 10 minutes of arc at a wavelength of 21 cm. This is comparable to the total angular size of a galaxy at a distance of 10 Mpc. That means that single-dish observations can derive detailed rotation curves only for very nearby galaxies – galaxies closer than two or three Mpc, such as M31. But, as we can see from eq. 4.1, radio interferometers, an array of dishes extending for 1.5 kilometers or more, can resolve structure on angular scales down to one-half of one minute of arc. The resolution of an interferometer is sufficient to map the distribution and motion of neutral hydrogen in many galaxies, not just the nearest. A single-dish telescope is sensitive to extended gas structures, but the interferometer can resolve details. Radio interferometers, like those at Owens Valley and Westerbork, can determine the rotation curves for a much larger sample of spiral galaxies.

In the early 1970s David Rogstad and his student, Seth Shostak, observed several spiral galaxies using the Owens Valley interferometer, and their initial results were summarized in a paper written in 1972. For five spiral galaxies observed with sufficient resolution to determine the rotation curves, they found that the rotation velocity rose sharply with radius to a maximum value and then remained constant at that value beyond the visual object. Referring to Freeman's law that the light distribution in spiral galaxies falls exponentially, Rogstad and Shostak, in a somewhat understated conclusion, "confirm the requirement for low-luminosity material in the outer regions of these galaxies". In other words, spiral galaxies appear to possess *darker* matter in the outer regions of the visible disk.

One of these first relatively high-resolution interferometer maps of the distribution and motion of hydrogen was that of NGC 2403, a spiral galaxy with a very ordered appearance at a distance of 3.6 Mpc. Apart from high resolution, a second advantage of an interferometer is that such an instrument, in a single pointing, produces a two-dimensional image of the 21-cm line emission from the galaxy as in Fig. 4.3; interferometers observe the radial-velocity field of the neutral hydrogen over the entire image of the galaxy, down to the resolution of the instrument. Single-dish telescopes can also map the two-dimensional velocity field for nearby galaxies, but this requires many pointings, i.e., many separate observations, to cover the galaxy. The same is true with optical spectroscopic observations which usually observe a one-dimensional velocity curve as determined by a single slit positioned along the major axis.

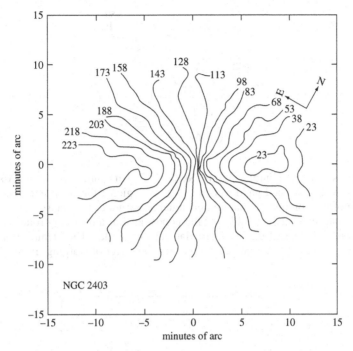

Fig. 4.8. Isovelocity contour map for NGC 2403. Each curve shows the fixed observed radial velocity of the gas as observed by the interferometer at Owens Valley. This pattern is typical of an inclined gas disk in circular motion about the center of the galaxy as in the hypothetical case shown in Fig. 4.3. From Shostak (1973).

The two-dimensional velocity field is most often illustrated by way of an isovelocity contour map like that shown in Fig. 4.3, and such a "spider" diagram is reproduced in Fig. 4.8 for NGC 2403 from the observations of Rogstad and Shostak. Each curve shows the locus of constant radial velocity, and this is the pattern that would be produced by a gas lying in a single plane inclined to the line-of-sight and executing simple rotation about the center of the galaxy (remember, only the component of the gas velocity toward or away from the observer is seen). An advantage of this sort of diagram is that distortions from circular motion in a plane are immediately apparent. For example, a warping of the gas plane is obvious in such a display because it also produces a characteristic warping of the velocity field as is shown in Fig. 4.9; the two-dimensional velocity field for a warped galaxy NGC 5055 observed later at Westerbork by Albert Bosma (1978). If the radio astronomer wishes to determine a precise rotation curve, then this warping must be taken into account.

Rogstad had earlier worked out how to correct for this effect. He took the gas layer of the galaxy to be represented by a system of rings. Then the velocity field

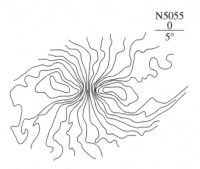

Fig. 4.9. Isovelocity contour map of NGC 5055. This pattern is characteristic of a galaxy with a warped gas layer in the outer regions; the warp is evident as a twisting of the contours. A one-dimensional spectrum of this galaxy taken along the apparent major axis would give the appearance of a declining rotation curve even though the actual curve of circular velocity is flat. This demonstrates the importance of identifying and correcting for the effect of warping. From Bosma (1978).

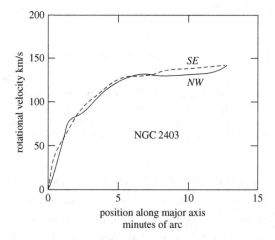

Fig. 4.10. The rotation curve of NGC 2403 measured by Shostak and Rogstad. This is one of the first neutral hydrogen rotation curves derived from radio synthesis observations with the Caltech interferometer at Owens Valley. This form, rising to a constant value and showing no decline, is typical of spiral galaxy rotation curves. The surface brightness distribution of this galaxy has the characteristic exponential form with a radial scale length a bit larger than 2 minutes of arc. Thus the rotation curve extends to five radial scale lengths, well beyond the point where the rotation curve should exhibit a Keplerian decrease if light traces mass. From Shostak (1973).

of the galaxy could be matched by letting each ring have its own separate inclination, but in fact, NGC 2403 is a fairly well-behaved galaxy and does not need much correction. For this galaxy, Shostak derived the rotation curve shown in Fig. 4.10. The rotation curve rises to a maximum but then does not decline out to the observed extent of the gas disk. Significantly, *the rotational velocity does not exhibit a Keplerian decline beyond the bright inner regions of the disk.*

This rotation curve, published by Shostak in 1973, was met with a great deal of skepticism, as were the single-dish results of Roberts and Whitehurst on the extended rotation curve of M31. Many astronomers just did not believe that this could be possible; the general feeling was that all instrumental and systematic effects had not been correctly taken into account. Radio telescopes, both single-dish and interferometers, have a complicated beam pattern. The image of a point source is not just a circular pattern, but something rather more intricate; emission can be seen well beyond the central beam, sometimes many arc minutes beyond. So the work of Shostak and Rogstad (1973), as well as that of Roberts and Whitehurst, was dismissed by some as an effect of poorly understood beam shapes. In fact, such criticism was unfair; the radio astronomers understood their antennae and receivers very well indeed and had taken care to correct for instrumental effects.

There was a great reluctance to accept the idea that rotation curves could be flat beyond the bright disk. The early models of galaxy disks had Keplerian decline built into the rotation curve; and observers insisted upon using these models in spite of evidence to the contrary. The earliest advocate and spokesman for the reality of flat rotation curves was certainly Morton Roberts. But in every talk he gave on the subject, at every conference or colloquium, he met considerable opposition. It was argued that he did not fully understand instrumental effects or the systematic uncertainties in the galaxy itself (e.g., warping, non-circular gas motion). These early observations of flat rotation curves were often ignored or dismissed, but by 1980, opinion had shifted dramatically.

4.4 Flat rotation curves: perception approaches reality

In 1975 there remained considerable doubt about the reality of flat rotation curves; however, in the second half of the decade, there were two developments which changed this perception. The first was that 21-cm line receivers had become available for the radio interferometer – the Westerbork synthesis radio telescope, or WSRT – and the neutral hydrogen was being mapped in a number of spiral galaxies. This was significant because, the WSRT was, at that time, the best radio interferometer in the world, with 12 dishes, 10 fixed and two movable, in an east–west line of 1.5 km. Because of this relatively long baseline and large collecting area, it combined high resolution (25 arc seconds at the celestial pole) with high sensitivity.

The second development was the systematic observation of rotation curves of spiral galaxies at optical wavelengths. Gas-rich spiral galaxies contain many so-called HII regions, that is, regions of ionized gas surrounding young stars. The young stars emit ultraviolet radiation which ionizes the surrounding gas. The ionized gas emits conspicuous spectral lines, arising primarily from recombining hydrogen, in the visible part of the electromagnetic spectrum. So, just as the rotation curve can be measured in the radio 21-cm line, the rotation curve can also be measured in the optical emission lines of ionized gas. This has an advantage over 21-cm line observations, as well as a disadvantage. The advantage is that the rotation curve can be measured with high spatial resolution along the major axis of the galaxy; detailed structure in the rotation curve can be seen. The disadvantage is that, by definition, the rotation curve measured with visual emission lines cannot be determined beyond the optical disk of the galaxy because these lines are only seen where there are stars.

In any case, by the late 70s high-resolution rotation curves were becoming available at both radio and optical wavelengths. At Westerbork, a number of galaxies were observed and catalogued by a student, Albert Bosma, as part of his PhD project. In his dissertation, which appeared in 1978, he presented the complete velocity fields (not just rotation curves, but the two-dimensional radial-velocity maps – the spider diagrams) of 25 galaxies, and, in most cases, the rotation velocity was measured beyond the optical image of the galaxy. Because of the two-dimensional velocity fields Bosma could identify systematic distortions from pure circular motion. He was able to classify the various kinds of distortions: bars which cause deviations from circular motion in the inner regions and warping of the gas layer which distorted the velocity field in the outer regions. In particular, he extended the method of Rogstad to correct for the effect of warping (in addition to the ring inclination, Bosma allowed the rotation velocity of each tilted ring to be a free parameter in fitting to the two-dimensional radial-velocity field). He also developed his own technique for dealing with non-circular motion due to deviations from axial symmetry.

Bosma found that the rotation velocity never declined in the Keplerian way, but in most cases remained constant beyond the optical disk. This is illustrated in Fig. 4.11 which is reproduced from Bosma's thesis. This collection of rotation curves is somewhat inhomogeneous; not all were observed by Bosma at Westerbork. But those objects with extended HI distributions demonstrate clearly that the rotational velocity remains constant well beyond the visible disk. The implication is that the total enclosed mass appears to be growing with radius while the total enclosed luminosity is not. Thus Bosma demonstrated that the local mass-to-light ratio increases dramatically in the outer parts of many spiral galaxies (not just one or two), as in the hypothetical case shown in Fig. 4.7.

At the same time, using the large optical telescopes of the Kitt Peak National Observatory near Tucson, Arizona, Vera Rubin, Kent Ford and Norbert Thonnard were making precise spectroscopic observations of the rotation curves of spiral

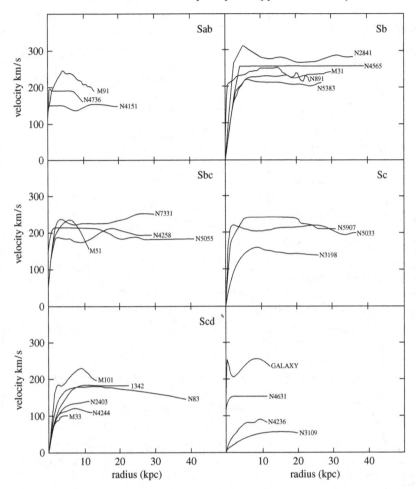

Fig. 4.11. The rotation curves of 25 spiral galaxies determined from 21-cm neutral hydrogen observations. From the PhD dissertation of Albert Bosma in 1978.

galaxies using the visible emission lines of hydrogen and nitrogen. An example of what they saw is shown in Fig. 4.12.

The rotation curve of UGC 2885 is especially interesting. This is a very large bright spiral galaxy with HII regions extending beyond 80 kpc (if $H_0 = 50$ km/s/Mpc). In this figure, the rotation velocity is roughly constant to the limit of the observations with no suggestion of a Keplerian decline. These results *appeared* to be completely consistent with the extended rotation curves in 21 cm of neutral hydrogen observed by Bosma at Westerbork WSRT and were perceived by many astronomers as even more convincing evidence for an increase in the mass-to-light ratio in the outer regions of spiral galaxies. But, in fact, such a direct

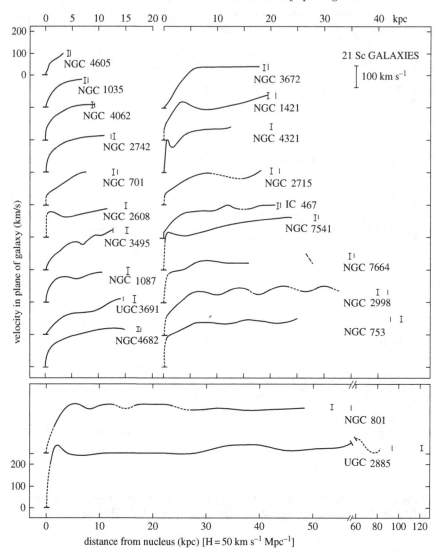

Fig. 4.12. The optically determined rotation curves of 21 spiral galaxies by Rubin, Ford and Thonnard (1980).

comparison of the radio and optically derived rotation curves, as in Figs. 4.11 and 4.12, gives a somewhat false impression. In Fig. 4.13 I plot the optically derived rotation velocity of UGC 2885 observed by Rubin, Ford and Thonnard and the 21-cm line rotation velocity of NGC 2403 observed by Begeman (1989) against the radial distance to the center, not in physical units (kpc), but in terms of the optical size of each galaxy (Begeman's later observations essentially confirmed those of Rogstad and Shostak but extended somewhat further). Here we get quite

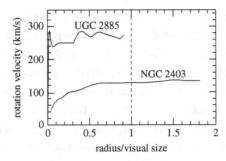

Fig. 4.13. The rotation curves of UGC 2885 (optical) and of NGC 2403 (radio) plotted in terms of a radius scaled to the visible size of each galaxy. The visible radius is taken to be that point at which the surface brightness falls to a certain faint level (25 magnitudes per square arc second, or, roughly 5 solar luminosities per square parsec).

a different impression about which rotation curve is more extended. The 21-cm line rotation curve extends to almost twice the optical radius of the galaxy, defined here as the distance at which the surface brightness reaches a particular faint level (25 magnitudes per square second of arc); there is no hint of a decline. The optically derived rotation curve, however, can be derived only within the bright image of the galaxy, i.e., within the obviously visible mass distribution of the galaxy. Although the point was not appreciated fully at the time, the radio observations clearly make a stronger statement about the necessity of dark matter in the outer parts of spiral galaxies (I will discuss this point further in the following chapter).

In any case, it is difficult now to appreciate the impact of these observations of Bosma and of Rubin and her collaborators. In a period of two or three years (1978–1981), the phenomenology of flat rotation curves went from being a somewhat dubious result of radio astronomers to an increasingly accepted view of spiral-galaxy kinematics. There seemed to be little doubt anymore – galaxy rotation curves did not decline in a Keplerian manner. The rotational velocity appeared to rise to a maximum at a radius of a few kiloparsecs, and then remain at this constant value even, as demonstrated by 21-cm line observations, well beyond the optical disk. Moreover, the perception of what spiral galaxies are changed dramatically. The observations were completely consistent with the earlier suggestion of Ostriker and Peebles that disk galaxies which were apparently rotationally supported required a massive spheroidal halo to stabilize the disk. The dark halo extended beyond the visible disk and gave the appearance of an increasing mass-to-light ratio in the outer regions.

The measurement of extended flat rotation curves was simultaneous with, but independent of, the theoretical considerations leading to the proposal of dark halos. Of course, given the publication of the Ostriker and Peebles proposal in 1973 and the almost simultaneous work on neutral hydrogen rotation curves by Rogstad and Shostak and by Roberts and Whitehurst, the theorists and observers were aware of, and no doubt influenced by, each other. But while the theoreticians certainly took some comfort in the observations of extended flat rotation curves, their proposal did not result from actual observations of galaxies but from theoretical considerations of the stability of disk galaxies. At the same time the observers were not driven by any need to test the proposal of dark halos; they were taking advantage of the new technology – sensitive receivers and radio interferometers – to explore the kinematics of the outer regions of spiral galaxies. As it turned out, the observations were entirely consistent with the idea that the visible spiral galaxy is only one component of the actual system – that the visible disks are immersed in enormous dark halos that extend far beyond the disk. However, taken by themselves (and this is an important point), the observations were also consistent with disks that became darker and darker in the outer regions; there was no observational requirement that the dark component should be distributed in a spheroidal system supported by random motion.

In any case, theory and observations naturally raised profound questions. What is the dark matter from which these dark halos are constructed? Is the dark matter problem of galaxies related to the dark matter problem of galaxy clusters found years earlier by Zwicky? It would certainly seem to be an efficient explanation of both phenomena if the dark matter were the same in these two environments. Simultaneous with the observations of flat rotation curves, there was an increasing awareness that something was missing on cosmological scales: it does not seem possible to form galaxies and clusters and superclusters without some additional assistance from the gravity of an unseen component (Chapter 6). Are cosmological and galactic requirements for dark matter all part of the same problem?

The early speculation on the composition of the dark matter content of galaxies centered upon low-luminosity stellar objects – low-mass, or "failed stars", or stellar remnants like white dwarfs or neutron stars or even black holes. Such dark matter is comprised of ordinary atoms and ions, but most of the mass is in the form of protons and neutrons, i.e., baryons. For that reason this sort of dark matter is called "baryonic dark matter". But there is another possibility – a more exotic form of dark matter known as "non-baryonic dark matter".

Before taking up the cosmological requirements for dark matter and the necessary properties of the hypothetical particles, I will describe a counter-revolution – compelling arguments against the necessity of dark matter on galaxy scales. In fact, these came down to arguments that questioned optically derived rotation curves as evidence for dark matter – arguments which led to important insights into the nature of the discrepancy in galaxies.

5

The maximum-disk: light traces mass

5.1 Reaction follows revolution

In the summer of 1982 Symposium number 100 of the International Astronomical Union took place in the French pre-alps, in the university town of Besançon. The subject of the symposium was the "Internal kinematics and dynamics of galaxies". By this point, the existence of a discrepancy between the dynamical mass and the visible, or detectable mass of galaxies had become a generally accepted fact. This acceptance was based largely upon the observations of the extended and non-declining rotation curves of spiral galaxies. In particular, the optical rotation curves had played an extremely important role in this perception. Observations, primarily by Rubin and collaborators, in optical emission lines of the rotation curves of very large galaxies such as UGC 2885 (Fig. 4.12) were especially important in enforcing this perception, even though, by definition, optical rotation curves could not extend beyond the visible disk of the galaxy.

The commonly believed explanation for this discrepancy was that already provided by Ostriker and collaborators: that spiral galaxies are immersed in massive and extended dark halos, dark halos with a mass that only increases with scale. In the previous decade there had been considerable resistance to this view, but by the time of this IAU Symposium, it was well on the way to becoming the central paradigm of galaxy structure and dynamics. There were still, however, pockets of resistance to this point of view.

One of the most effective counter-arguments was presented by Agris Kalnajs at this meeting. Kalnajs was a theoretician working at the Mt. Stromlo Observatory in Australia, but before going into his argument, we return to a bit of history. In 1970, Ken Freeman had discovered that the light distribution of disk galaxies could be generally described by an exponential curve with a central surface brightness that, unlike the size scale, did not vary much from galaxy to galaxy. Freeman had calculated the rotation curve of these exponential disks and found that beyond three or four scale lengths the rotation curve should decline in an almost Keplerian way. This exponential curve is quite a good approximation in many cases, but, of course, there are deviations from this idealized form (the most conspicuous deviation is, in a number of cases, a separate central spheroidal component called a bulge).

In 1981 Piet van der Kruit at the Kapteyn Institute along with Leonard Searle of the Mount Wilson Observatory (Carnegie Institution) found that often the starlight of the disks does not just continue to fade exponentially but, in fact, seems to have an edge, a rather abrupt truncation, at about four scale lengths. So it seemed appropriate to calculate the rotation curve of truncated exponential disks. This was done in 1982 by Stefano Casertano, then a student working at the Kapteyn Institute and later at the Institute for Advanced Study in Princeton. He found that, before the truncation, the rotation curve in a thin disk did not decline so rapidly but, in fact, seemed rather flat. This is shown in Fig. 5.1, where we see the effect on the rotation curve of truncating the disk at various radii given in units of the exponential scale. This suggests that for a realistic mass distribution in a disk, a flat rotation curve, at least within the disk, does not necessarily require the presence of dark matter.

Thirty years earlier Martin Schwarzschild had calculated the rotation curve of M31 assuming that the stellar mass was distributed in a thin disk and that light traces mass perfectly, i.e., that the mass-to-light ratio in the galaxy is constant (Chapter 2). He found an acceptable match between his calculated rotation curve and the optically observed rotation curve (by Mayall) and concluded that there was no evidence for an increasing M/L in the outer regions of M31.

Does light trace mass? With a vastly improved and enlarged set of data on galaxy kinematics and photometry, this was the question revisited by Kalnajs in 1982. He was unimpressed by the evidence in favor of dark halos, especially that evidence

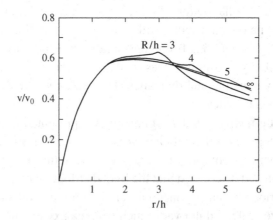

Fig. 5.1. The rotation curves of exponential disks truncated at various radii in units of the exponential scale length, calculated by Casertano (1983).

based upon optical rotation curves, and he also doubted the Ostriker–Peebles stability criterion that seemed to require a contribution of mass from the dark halo comparable to that of the disk within the visible galaxy. Kalnajs reasoned that tracers of the gravitational force, such as emission lines from ionized hydrogen regions that are still embedded in the disk, would not be expected to exhibit a Keplerian decline of the rotation curve – independent of the presence of a halo. Kalnajs also realized that each galaxy has its own unique surface-brightness distribution and that the exponential form is only a general approximation. In any individual spiral galaxy one should use the actual distribution of visible starlight in that particular disk to estimate what the mass distribution should be. And then, as did Schwarzschild years before, one could compare the predicted rotation curve with the observed rotation curve before making claims about the necessity of dark matter or a radially increasing mass-to-light ratio.

And this is what Kalnajs did. He considered four galaxies for which there were accurate observations of the radial distribution of starlight – the surface brightness of the disks as a function of radius. He assumed that the surface brightness of the disk is a perfect tracer of the stellar mass surface density in the disk – that the mass-to-light ratio of the stellar disk does not change with radius, so the form of the mass distribution is directly observed. Then, further assuming that this mass really is distributed in a thin disk and not in a spherical object, he could calculate the Newtonian gravitational force and therefore, the rotation curve. For these four galaxies the rotation curve in optical emission lines had been measured (two of them by Rubin and collaborators), so he could compare the calculated rotation curve – calculated from the observed light distribution – with the observed rotation curve.

His results are shown in Fig. 5.2. Here Kalnajs has placed as much mass as possible into the stellar disks; for more massive disks the predicted curve would exceed the observed rotation curve. This assumption would later become known as the "maximum-disk approximation". We can see that there is a remarkably precise agreement between the calculated and observed rotation curves. When Kalnajs presented this figure at the symposium it caused quite a stir. It shook the recently acquired view of many people that dark matter was necessary to explain non-declining rotation curves. The usual interpretation of the observations, especially the optical observations – that non-declining rotation curves imply dark matter – was very strongly challenged.

In these plots there is one free, or adjustable, parameter, and that is the mass-to-light ratio of the stellar component of the disk. These numbers are given in the figure captions, and the values of 5.0 or 6.5 in solar units are somewhat higher than that expected of "normal" stellar populations, but certainly not high enough to require a massive dark halo contributing substantially to the total mass within the

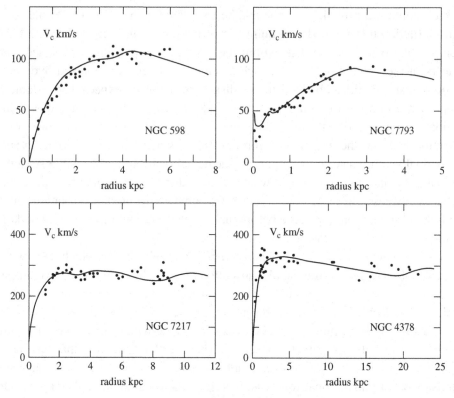

Fig. 5.2. The rotation curves (solid curves) of four spiral galaxies calculated by Kalnajs assuming that the light exactly traces the mass distribution and that the mass is in a thin disk. The points show rotation curves determined from observations of optical emission lines; thus, the measured rotation curves do not extend beyond the visible disk. The adopted M/L values for the stellar disks are 5.0, 2.9, 4.2 and 6.5. From Kalnajs (1983).

visible disk. If such an analysis had been done several years earlier, it is doubtful that the concept of dark matter would have been so readily accepted.

Of course, these were only four galaxies, and it could always be argued that Rubin and collaborators had observed many more galaxies (about 10 times more) having never seen a declining rotation curve; this surely amounted to compelling evidence of an extended dark matter distribution. But then a few years later (1986), Steve Kent of the Harvard Smithsonian Center for Astrophysics, repeated Kalnajs analysis for 37 spiral galaxies from the sample of Rubin *et al*. He made his own photometric observations tracing the distribution of visible light in these galaxies, and, following Kalnajs, assumed that the light is a perfect tracer of the mass. There was one complication: in a number of these galaxies there is clear evidence for a central spheroidal component in the light distribution – a bulge. It was necessary for Kent to devise a method of separating the bulge from the disk – so-called

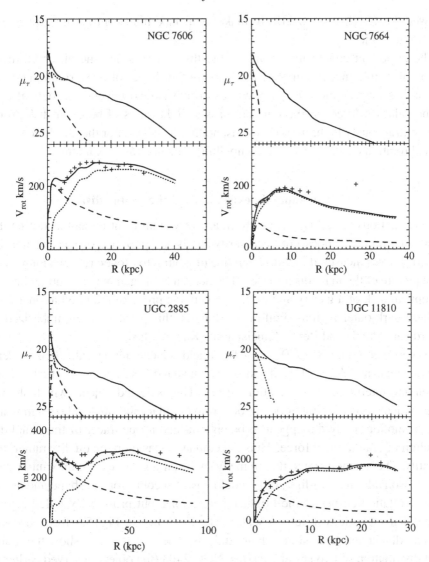

Fig. 5.3. The radial-light distribution and calculated rotation curves (solid curves) for four galaxies from the sample of Kent (1986). The points are observed rotation curves by Rubin and collaborators (1980). The fitted M/L values for the disk are 3.2, 1.6, 5.5 and 4.1. The photometry of the galaxies also implied an inner-bulge component with comparable M/L ratios.

bulge-disk decomposition. Because bulges are observed to have different colors than disks, it is also permitted to let the bulge assume a different mass-to-light ratio than the disk. This adds a second free parameter to the rotation-curve fits for many of these objects, but the procedure is basically the same as that adopted by Kalnajs. A sample of Kent's results are shown in Fig. 5.3, where again Kent has allowed the visible components to make their maximum contribution to the total mass of the object.

The important conclusion to draw from this figure is that the observed distribution of visible matter can generally explain the shapes of these rotation curves, at least, as long as those rotation curves do not extend beyond the optical edge of the galaxy. Moreover, the implied mass-to-light ratios of bulge or disk do not seem outrageously high. In other words, neither the shape nor the amplitude of the optical rotation curves constitute compelling evidence for dark matter.

5.2 The anomaly exists beyond the visible disk

The calculations of Kalnajs and of Kent are very significant in demonstrating that a flat rotation curve does not necessarily imply the presence of dark matter, but rotation curves measured in the 21-cm line of neutral hydrogen reveal a more complete picture of the mass discrepancy. The neutral hydrogen may extend far beyond the optical disk of a galaxy and often is a genuine probe of the gravitational force in these dark outer regions. Following Roberts, this point was re-emphasized by Tjeerd van Albada and Renzo Sancisi at the Kapteyn Institute.

In work beginning in 1980, van Albada and Sancisi, along with their students Kor Begeman and Adrick Broeils, took a somewhat different approach to the observation and interpretation of rotation curves. They selected objects which showed little evidence for warping or deviations from circular motions – galaxies in which the gas motion really did appear to be circular and a true tracer of the radial distribution of gravitational force. Thus the emphasis was more on quality rather than quantity. They demonstrated that in many such spiral galaxies, the maximum-disk, with reasonable mass-to-light ratio, could easily account for the shape and amplitude of rotation curves in the bright inner regions, but profoundly failed beyond a few characteristic disk length scales. This is shown in Fig. 5.4 from the work of van Albada and Sancisi and their students. The panels above show the radial-light distribution of two spiral galaxies, NGC 2403 (the object observed earlier by Rogstad and Shostak) and NGC 3198 (previously observed by Bosma), and below, the solid curves are the rotation curves calculated from this light distribution with the usual light-traces-mass assumption. The points with error bars are the rotation curves inferred from updated 21-cm line observations at Westerbork WSRT – observations which extend well beyond the bright inner disk.

The calculated rotation curves are again based upon the maximum-disk assumption; this sets the amplitude so that the observed rotation velocity in the inner regions is completely explained by the visible matter as in the calculations of Kalnajs and Kent. We see that the calculated rotation curves fit the observed curves in the inner region, but then decline in an almost Keplerian fashion beyond. The

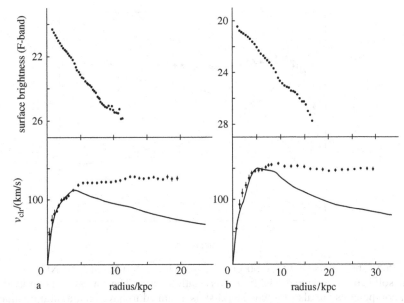

Fig. 5.4. The radial-light distribution and calculated rotation curves (solid curves) for two spiral galaxies shown by van Albada and Sancisi (1986) assuming that light traces mass. The points are the rotation curve observed in the 21-cm line of neutral hydrogen, which extends far beyond the bright inner disks of these two galaxies.

observed curves, however, remain essentially flat beyond the peak. This demonstrates a very serious discrepancy in the outer regions. If we construct a mass model to fit the observed rotation curve, then we find that the local dynamical mass-to-light ratio in the outer regions is in excess of 1000. These 21-cm observations do constitute compelling evidence for a discrepancy – or, viewed in terms of Newtonian dynamics, for missing luminous mass.

The observations do not require, however, that the dark mass is distributed in a dark spheroidal component. The flat rotation curves could just as easily result from a disk which becomes systematically darker in the outer region. The disk must be pressure supported, not rotationally supported, to overcome the stability problems pointed out by Ostriker and Peebles, but it is, in principle, possible.

The apparent success of the maximum-disk model raises another issue, however, which the paradigm of dark matter halos must yet confront: the so-called disk–halo conspiracy. This is a point made graphically clear by interpretation of the 21-cm line observations of the galaxy NGC 3198 published by van Albada, Bahcall, Begeman and Sancisi in 1985. In Fig. 5.5 we see the observed rotation curve of this galaxy and the rotation curve calculated from a model consisting of the maximum-disk (taken to be purely exponential here), and a spherical dark halo with a density distribution that falls like $1/r^2$ (or a surface density falling as $1/r$) in

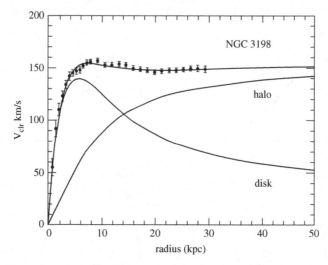

Fig. 5.5. The points show the observed 21-cm line rotation curve of NGC 3198. The solid curves are the model fit to this rotation curve. The curves of the individual components are also shown. The disk is assumed to have its maximum value consistent with the observed rotation curve in the inner regions (the "maximum-disk" assumption). The appearance of a flat rotation curve requires a careful coupling of the disk and halo curves; as the disk rotation curve falls, the halo rotation curve rises. From van Albada *et al.* 1985.

the outer regions. This density law provides an enclosed mass that grows as radius and, therefore, a rotation curve that is flat. But we also see that to produce a rotation curve that is flat overall, the halo rotation curve must be closely tuned to the disk rotation curve: the halo rotation curve must rise as the disk rotation curve falls.

This fine-tuning is known as the disk–halo conspiracy. If the disk dominates the mass distribution in the inner regions and the halo in the outer regions, then these two components must somehow conspire in order to produce a rotation curve that is so relatively flat and featureless from small to large radii. If the halo would completely dominate, then there would be no problem – it takes two to conspire. This is a problem that I will return to, especially in Chapters 8 and 10.

This analysis of the 21-cm line rotation curve of NGC 3198 provided the strongest case for the reality of the discrepancy on galaxy scales. Because of the clear interpretation of the gas motion beyond the disk as pure rotation about the center of the galaxy and because this gas disk extends to more than 10 exponential scale lengths (well beyond the point where the rotation curve should be Keplerian if light traced mass), there could no longer be any doubt that the Newtonian dynamical mass greatly exceeds the visible mass in the outer regions of spiral galaxies. But the contribution of dark matter, a dark halo, within the visible disk remained an open question. The principal assumption of the maximum-disk is that this halo contribution is minimal, but, that is an assumption. With respect to fitting rotation curves, a range of models is most often possible: from maximum-disk to no disk. But now, after 25 years, the role of the maximum-disk is more evident.

5.3 Low-surface-brightness galaxies and sub-maximal disks

How generally applicable is the maximum-disk approximation? It does seem to work well, particularly in the galaxies observed by Rubin and collaborators. The maximum-disk explains the shape of the rotation curve, but, in a number of cases, in order to explain the amplitude of the curve the required mass-to-light ratio is rather high compared to expectations from pure populations of stars with a normal distribution of mass. The Rubin *et al.* galaxies all have a high surface brightness, but, in the last 20 years, a large number of low-surface-brightness galaxies have been discovered. And this population of low-surface-brightness galaxies has clarified the role and applicability of the maximum-disk.

According to Freeman, the central surface brightness of a spiral galaxy disk, I_0 in eq. 4.7, appears to be roughly constant from galaxy to galaxy, but the scale length, h, varies. In other words, all spiral galaxies seemed to have about the same central surface brightness, but they vary in size: the big ones have high luminosity and the small ones have low luminosity. This became known as Freeman's law.

The aspect of a characteristic central surface brightness now seems to be somewhat of a simplification. In 1976, Mike Disney, then at the Kapteyn Institute, pointed out that Freeman's characteristic surface brightness could actually be due to a selection effect: fainter galaxies would be difficult to detect because their surface brightness would be below that of the sky and brighter galaxies would be confused with stars. In 1979 Ronald Allen and Frank Shu at Berkeley demonstrated that this selection actually works only in one direction: that there is indeed an upper limit to the surface brightness of galaxies but not necessarily a lower limit. In other words, there could be a large population of faint galaxies but no galaxies with a higher surface brightness. Freeman's characteristic surface brightness is, in fact, an upper limit. By 1990 a large number of "low-surface-brightness" (LSB) galaxies had been discovered and catalogued. I stress that I am not talking about the apparent magnitude or the intrinsic luminosity of a galaxy, but about the surface brightness. If L is the luminosity or total power of visible light emitted by a galaxy, then the surface brightness would be more like L/h^2 where h is the exponential scale length of the galaxy. In high-surface-brightness galaxies, the usual objects with optically derived rotation curves, this average surface brightness is in excess of 100 $L_\odot/pc^2$. For low-surface-brightness galaxies this may be 10 times smaller.

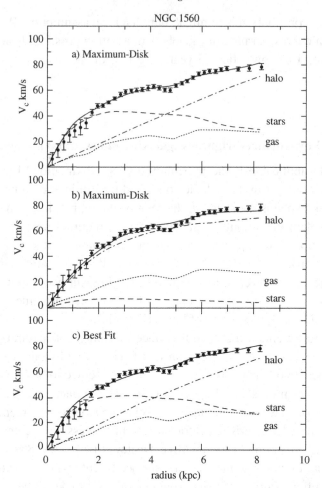

Fig. 5.6. The points show a 21-cm line rotation curve of NGC 1560, a galaxy with low surface brightness, observed by Adrick Broeils (1992). The top panel shows a maximum-disk model fit, the middle panel is a model in which the halo contributes almost all of the mass and the lower panel is the best model fit. The maximum-disk fails to reproduce the shape of the rotation curve within the optical disk. Moreover, the M/L of the maximum-disk is almost 5, extremely high for such a small blue galaxy.

In the early 1990s observers began to measure the rotation curves of LSB galaxies, and an interesting trend emerged. These objects apparently do have a large discrepancy between the visible and dynamical mass well within the optical image of the disk – they are far from maximum-disk. This is shown in Fig. 5.6, where we see a low-surface-brightness galaxy (or LSB) observed at Westerbork (WSRT) by Adrick Broeils.

Here the maximum-disk fails to reproduce the shape of the rotation curve within the visible disk. Moreover, the mass-to-light ratio is almost five in solar units – exceedingly high for a blue-star-forming galaxy disk. Subsequent work with the WSRT by Stacy McGaugh, Erwin de Blok and Rob Swaters confirmed this trend: for LSB galaxies, there is a considerable discrepancy between the Newtonian dynamical mass and the visible mass within the optical image of the galaxy, or, with Newtonian dynamics, there must be a large contribution of dark matter within the inner regions of these galaxies. In Swaters dark-halo fits to the observed 21-cm line rotation curves of about 30 LSB galaxies (1999), the Newtonian rotation curve of the visible disk, assuming light traces mass, fails to reproduce the shape of the observed rotation curve and in several cases, requires mass-to-light ratios greater than 10 to come near to matching the amplitude. It is now clear that the maximum-disk is an ideal that is only approached in the galaxies with high surface brightness. This is a significant aspect of the discrepancy in galaxies and one which I will discuss in later chapters.

5.4 Reflections on observations of rotation curves

It is easy to identify the astronomer who first discovered the discrepancy between the dynamical and visible mass of a large astronomical system and introduced the concept of "dark matter" to explain this discrepancy. That is Fritz Zwicky (1933) in his work on the Coma cluster of galaxies. But it is not possible to single out an individual who discovered, by means of astronomical observations, the need for dark matter on a galaxy scale. Several astronomers made substantial contributions to this realization.

The strong case for dark matter beyond the visible disk of spiral galaxies was present in the early 1970s in the 21-cm line observations. This was evident in the radio interferometric work of Shostak and Rogstad and in the single-dish observations of Roberts and Whitehurst. In particular, Morton Roberts deserves credit for being the first to grasp the consequences and argue forcefully for the reality and importance of non-declining rotation curves. Albert Bosma's encyclopedic thesis in 1978 had a major impact. His 21-cm line observations demonstrated that the discrepancy appeared to be present not just in one or two objects but is a general aspect of spiral galaxies. Bosma's work is in some sense under-appreciated because it contains almost too many interesting results – e.g., warped gas layers from kinematic evidence only, the pattern of non-circular motions due to bars and weak non-axisymmetric distortions – and in a sense presents a trees vs. forest problem of perception.

The galaxies observed by Rubin and collaborators were all, of necessity, high-surface-brightness galaxies for which, as we now know, the maximum-disk approximation with its underlying assumption that light traces mass appears to work well in explaining both the shape and the amplitude of the optically measured rotation curves. This was demonstrated by the work of Kalnajs and Kent, while van Albada and Sancisi and their collaborators established that the substantial evidence of a serious discrepancy between the dynamical and observable mass on galaxy scales comes from 21-cm line observations of the neutral hydrogen that extends far beyond the bright optical disk. The HI rotation curve of NGC 3198 published in 1985 by van Albada and collaborators (Fig. 5.5), extending more than 10 exponential disk scale lengths or three times further than the visible disk, presents compelling, indisputable evidence for a large observational anomaly – dark matter in the context of Newtonian gravity – in the outer regions of galaxies. The subsequent observations of low-surface-brightness galaxies, which reveal a serious discrepancy well within the optical disk, came some years later. The evidence for dark matter from the optical rotation curves of high-surface-brightness galaxies is not compelling.

Yet Rubin's observations had a powerful influence upon the perception that there is dark matter on galaxy scales. To many, the optically observed rotation curves appeared to definitively prove this. The observation of rotation curves which were flat from the bright inner regions to the dark exterior of spiral galaxies appeared to be consistent with the stability argument which required a dominant dark component even within the bright inner regions (no distinct signatures of separate halo and disk components). This remained a potent idea even though serious objections (primarily by Alar Toomre of MIT and Agris Kalnajs) had been raised against the precise criterion of Ostriker and Peebles.

More generally, the perceived importance of optical rotation curves is only one example of a broader social aspect of the practice of science; with respect to the dark matter problem there existed a group dynamic which first refused to accept the obvious evidence for a significant anomaly in galaxies and then, in a total reversal, saw dark matter everywhere, even underlying observations having a more conventional interpretation. It appears that science is not different from other human endeavors (politics, economics) in this respect; in a social context, the perception of reality has significance approaching that of the reality itself, at least temporarily. Ultimately, the practice of science is distinct from other social activities in that the technique developed for probing an objective reality is the final arbiter of truth.

6

Cosmology and the birth of astroparticle physics

By 1980 the perceived problems of the stability of rotationally supported disk galaxies and the observations of non-declining rotation curves of spiral galaxies had led most astronomers to accept the idea that galaxies are embedded in a dark halo that becomes dynamically more important in the outer regions. There were counter-arguments: Kalnajs, for example, disputed the stability criterion and demonstrated that rotation curves from optical emission lines presented no compelling evidence for dark matter (in so doing, he illuminated the nature of the discrepancy in galaxies, as we have seen). But most people active in this field accepted the presence of a discrepancy and the view that galaxies were darker in the outer regions. Astronomers in general thought in terms of rather conventional dark matter – cold gas, very low-mass stars, failed stars (or super planets), stellar remnants such as cold white dwarfs, neutron stars, or low-mass black holes – i.e., baryonic dark matter.

At about the same time a rather different idea was gaining credence among cosmologists and particle physicists: that the dark matter consists of subatomic particles; non-baryonic dark matter that interacts only weakly with baryons and photons. This was a highly unconventional idea at the time, but now the concept of non-baryonic dark matter has become the paradigm; very few astronomers or cosmologists invoke ordinary, baryonic dark matter. This paradigm shift emerged from cosmological considerations, primarily from the problem of structure formation in a nearly homogeneous Universe combined with observations of the cosmic microwave background radiation – the CMB. How this major shift in ideas developed is the subject of this chapter.

6.1 A brief history of modern cosmological models

The Big Bang model of the Universe has its roots in a solution to Einstein's gravitational field equations given by the Russian mathematician Alexander Friedmann in 1922. Friedmann demonstrated that, in the absence of a cosmological constant, Einsteinian cosmology cannot be static: the Universe is either expanding or contracting, and Friedmann reached this conclusion on purely theoretical grounds well before Hubble's discovery in 1929 that the Universe actually does appear to be expanding. Several years later, Georges Lamaître, drawing on Hubble's discovery, proposed a cosmology in which the Universe was originally much smaller and denser – a primeval atom – the explosive breakup of which formed the present chemical elements.

In the 1940s George Gamow (a student of Friedmann who had immigrated to the USA) put electromagnetic radiation into the mixture and pointed out that at some point in the early Universe, depending upon the present temperature of the radiation and the matter density, the mass density of radiation would dominate over that of ordinary matter (see eq. A6.5). Gamow went further: he proposed that all of the elements more massive than hydrogen could actually be synthesized in the early Universe when the temperature and density were high. In order to accomplish this the radiation had to be at a temperature on the order of one to 10 billion degrees, but with the subsequent expansion of the Universe the radiation would have cooled to a present temperature of only a few degrees. This proposal was published in 1948 in a famous paper with Gamow's student, Ralph Alpher, and the physicist Hans Bethe (taken on board apparently for the benefit of having a Greek alphabetic author list). Gamow and his colleagues predicted that a background radiation at a temperature of a few degrees kelvin should exist. It was quickly demonstrated (by Enrico Fermi, after hearing a colloquium given by Alpher at Columbia University) that heavy elements (heavier than the beryllium isotope with atomic mass 7) could not be produced in the early Universe, but the light elements (deuterium, helium, lithium) could indeed have their origin in this expanding hot fireball. Alpher then working with Robert Herman, predicted that, in order to accomplish this synthesis of the light elements, the present temperature of the background radiation should be about five kelvins (Alpher and Herman 1949).

At the same time (1946) a competing cosmology emerged – the "steady-state model" developed by the British astrophysicists Thomas Gold, Herman Bondi (1948) and Fred Hoyle (1948). This model is based upon the so-called perfect cosmological principle – the Universe appears the same not only to all observers at any point in space, but also to all observers at any point in time. Of course, the Universe is expanding, which would imply that the density of matter should decrease with time – in stark contrast to this principle. But Gold, Bondi and Hoyle proposed that new matter is continuously being created at a rate sufficient to keep the density constant. They argued that creation is a problem whether it is instantaneous (as in the Big Bang) or continuous.

An aspect of the steady state is that all elements heavier than hydrogen are created in the interiors of stars, and, indeed, there was convincing evidence that this is true. In spectroscopic observations of stars in the Milky Way, Martin and Barbara Schwarzschild (1950) demonstrated that old stars (stars not confined to the plane of the Galaxy) have a lower abundance of metals than do young stars – stars confined to the galactic plane and recently formed out of interstellar gas containing the "pollution" from nucleosynthesis in preceding generations of stars (for an astronomer, any element heavier than helium is a metal). The model gained further support from evidence that the oldest stars, in the system of globular star clusters surrounding the Milky Way, would be older than the Universe in the context of the Big Bang (based upon an incorrect determination of the Hubble constant). Finally a demonstration by Margaret and Geoffrey Burbidge along with William Fowler and Fred Hoyle that elements heavier than iron could be formed by neutron capture on iron nuclei in stars (1957) tipped the balance toward the steady-state model.

In the early 1960s the steady-state Universe appeared to be the emerging cosmological paradigm. But philosophically elegant though it was, there were cracks in its foundation. Hoyle himself (along with Roger Taylor) pointed out that if all helium were created in stars, then the total amount of energy radiated would be 10 times larger than the energy radiated by galaxies since their formation; the helium found in stars must be pre-galactic (1964). However, it was the discovery in 1965 of the three-degree cosmic background radiation, the CMB, by Arno Penzias and Robert Wilson – a prediction of Gamow's Big Bang – that effectively put an end to the steady-state model (although several of its advocates have fought a rear-guard action for decades). A significant consequence is that the Universe has a finite lifetime and a finite time in which to develop the wealth of structure we observe at present. (A more complete discussion of these early developments, in particular the substantial parallel contributions of Soviet scientists, can be found in the collection of essays "Finding the Big Bang" edited by James Peebles, Lyman Page and Bruce Partridge.)

James Peebles was one of the first of the modern (post-CMB) cosmologists. He had been a student of the famous experimental physicist Robert Dicke. He and Dicke along with P.G. Roll and Dave Wilkinson at Princeton were actually looking for the cosmic background radiation (1965) when Penzias and Wilson accidentally discovered it, and Peebles went on to make substantial contributions to the understanding of the thermal history of the Universe as outlined here in the Appendix (Section A6). He realized the importance of the epoch of decoupling – that the CMB photons that are detected all come from that opaque wall at a redshift of 1000. He determined how the predicted abundances of nuclei like helium and deuterium depend upon the present density of baryons, usually expressed as the baryonic fraction of the cosmological critical density, $\Omega_b = \rho_b/\rho_c$ (ρ_c is the average density required to close the Universe, eq. A5.6 see Peebles 1965, 1966, 1968). Ten years later (1974), following up on Peebles' work Richard Gott, James Gunn, David Schramm and Beatrice Tinsley pointed out that Ω_b must be considerably less than one in order to be consistent with the observed abundances of deuterium and helium; that if the density of the Universe is in fact equal to the preferred critical value, then something other than baryons must make up the difference.

Peebles was most interested in the formation of structure in the Universe and was the first to actually quantify the clustering of galaxies (before Peebles this subject was highly "astronomical", i.e., descriptive and qualitative). Following Gamow, he appreciated that, in a baryonic universe, structure formation via gravitational collapse could only begin at the epoch of decoupling ($z = 1000$). This makes the problem of structure formation even more severe; small positive density fluctuations in the otherwise smooth fluid of the Universe can only begin to collapse when the Universe is about 300 000 years.

6.2 Structure formation: dark matter again to the rescue

Structure in an almost smooth homogeneous universe can only form by gravitational instability if the size scale of density fluctuations, presumably generated during the very early inflationary epoch (at perhaps 10^{-34} seconds), exceeds the Jeans length; the distance traveled by a sound wave during a collapse timescale (see Section A8). The problem is that before decoupling of the photon and baryon fluids (before protons and electrons join to produce neutral hydrogen) the sound speed is comparable to the speed of light and, because the collapse time is roughly equal to the Hubble expansion time (the age of the Universe at a given epoch), the Jeans scale is comparable to a causally connected region – the horizon. Sub-horizon fluctuations in the photon–baryon fluid – fluctuations which could give rise to galaxies and clusters – do not grow but propagate as sound waves. The actual gravitational collapse of such fluctuations can only begin at decoupling (initially positive density fluctuations are not actually collapsing but are expanding – just more slowly than the Universe on average). In an expanding medium these fluctuations grow rather slowly – as a power law function of time rather than exponentially as in a static medium. So in order to produce the presently observed structure in the Universe, the original amplitude of the fluctuations at decoupling, the fractional variation in density ($\delta\rho/\rho$), must be relatively large ($\approx 10^{-4}$) and this should show up as comparable fluctuations in the CMB: the temperature of the CMB should vary by this fractional amount over angular scales of arc minutes. In the 25 years after the discovery of the CMB such small-scale variations of temperature were looked for but not seen. How then could structure form by the process of gravitational collapse?

By 1970, physicists had realized that, in the context of the Big Bang model, cosmological relic neutrinos are present in the Universe with a number density comparable to that of the CMB photons. The essential physics was outlined by the Soviet physicists, Yakov Zeldovich and Igor Novikov (see Zeldovich and Novikov, 1983): when the Universe is less than a few seconds old, the average energy of particles (the "temperature") exceeds two or three MeV and neutrinos are in thermal equilibrium with photons; i.e., there are weak interaction processes that convert neutrinos into photons and vice versa creating roughly equal numbers of both species. But as the Universe expands and the particle energy falls below 2 MeV (at an age of about one-half second) the timescale for these processes becomes longer than the age of the Universe – the neutrinos "freeze out" of the soup of photons and other particles (mostly electrons and positrons). Because neutrinos interact so rarely at these lower energies, they persist as relics of this very early phase of the Big Bang. In the Universe, at present, every cubic centimeter contains about 113 neutrinos of each type. Since there are three types of neutrinos this means more than 300 neutrinos per cubic centimeter.

Around 1970 there were several arguments that the neutrino might have a small but non-zero rest-mass. Such a possibility might, for example, solve the mystery of the missing solar neutrinos – a puzzle that was just beginning to emerge from the first neutrino telescope built by Ray Davis (a vast tank of cleaning fluid at the bottom of the Homestake gold mine in South Dakota). The difference between the predicted flux of neutrinos and that observed was immediately recognized as an anomaly by John Bahcall at the Institute for Advanced Study in Princeton – an anomaly possibly requiring new physics, as turned out to be the case (see Bahcall and Davis, 1976, for a discussion of these earlier results).

R. Cowsik and J. McClelland (1973) at the University of California in Berkeley, and, independently, A. Szalay and G. Marx (1976) in Budapest, realized that if neutrinos have even a small mass of a few eV (one electron volt is equivalent to 1.8×10^{-33} gram), then, because of the copious cosmological background of neutrinos, they could provide the missing cosmological mass in the Universe, a mass sufficient for $\Omega_0 = 1$. Following earlier work by Gershtein and Zeldovich (1966) they used the condition that $\Omega_0 \leq 1$ (the Universe does not seem to be closed) to place reasonable limits upon the mass of the neutrinos. The exact relation for three neutrino types is

$$\Omega_\nu = 0.02(m_1 + m_2 + m_3) \qquad (6.1)$$

where m_1, m_2 and m_3 are the masses of the three different neutrino types in eV (Ω_ν is the fraction of the closure density in neutrinos). This means that if all three types have masses of 17.5 eV then $\Omega_\nu = 1$; i.e., the Universe would consist mostly of neutrinos at the critical density. Both groups went on to speculate that neutrinos could be Zwicky's missing mass in the great clusters of galaxies. They did not discuss galaxy halos because at this early date the idea of concentrations of dark matter on a galaxy scale was not part of the general perception. This, in any case, seemed to be the first suggestion that dark matter in gravitationally bound astronomical systems might consist of non-baryonic subatomic particles.

The idea was taken further by a collaboration including James Gunn, B.W. Lee, I. Lerche, Dave Schramm and Gary Steigman (1978). They pointed out that any heavy stable particle like the neutrino that is a cosmological relic – a relic of the hot Big Bang – could comprise the dark matter. Their work enlarged the concept of non-baryonic dark matter to include hypothetical undiscovered particles more massive than the neutrino was thought to be. They also realized that such a relic particle would have major consequences for structure formation, and, although it was not fully appreciated at the time, for the unobserved fluctuations in the CMB temperature. If, in addition to baryons and photons, there exists another fluid – a fluid that dominates the matter budget of the Universe but does not couple to the photons – then the sound speed in this fluid may be much less than the speed of light and the Jeans length smaller than a horizon. Sub-horizon fluctuations in this decoupled fluid do not propagate as sound waves but continue to grow before decoupling while their counterparts in the baryon–photon fluid oscillate. At decoupling the baryons, released by the photons, fall into the potential wells established by the non-interacting fluid. In other words, structure can begin to form before decoupling of photons and baryons at $z \approx 1000$.

This possibility is illustrated in Fig. 6.1 for the case of hypothetical neutrinos with a mass of 17.5 eV (implying that $\Omega_\nu = 1$). This plot shows the total mass enclosed within a horizon and the masses enclosed within the Jeans length (the minimum unstable mass) for the neutrino fluid and the baryons as a function of redshift. As long as the neutrinos are relativistic – until their mean kinetic energy falls below the rest-mass energy of 17.5 eV – the effective speed of sound for the neutrino fluid is $c/\sqrt{3}$ just as for the baryon–photon fluid. But when the neutrinos cool below 17.5 eV, corresponding to a redshift of about 100 000, then the neutrinos continue to cool like a non-relativistic gas and the sound speed decreases; hence, the Jeans mass decreases below the horizon mass. When matter begins to dominate the density budget of the Universe, at a redshift of about 20 000, any surviving fluctuations in the neutrino fluid can begin to grow by gravitational instability while the baryon–photon fluctuations are still oscillating sound waves. At the decoupling of baryons and photons ($z \approx 1000$) the sound speed in the baryon fluid drops dramatically, as does the Jeans mass. For heavy particles, of the sort proposed by Gunn and collaborators, the mechanism would be basically the same, but heavier particles would be non-relativistic earlier, at a higher redshift.

This process is relevant to the constraints imposed by the absence of observed spatial fluctuations in the temperature of the CMB. By 1980 these constraints were

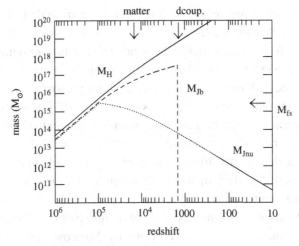

Fig. 6.1. The solid curve is the total mass within a horizon as a function of redshift for a hypothetical neutrino-dominated universe (three neutrino types each with a mass of about 17.5 eV and assuming $h = 0.75$). The dashed curve is the mass of baryon–photon fluid enclosed within one Jeans length for this fluid, and the dotted curve is the same for the neutrinos. The significant point is that the Jeans mass for the neutrinos begins to drop below the horizon mass when the neutrinos become non-relativistic at a redshift of about 100 000. Then structure can begin to form in the neutrino fluid well before decoupling. The baryons decouple from photons at a redshift of about 1000 (when neutral hydrogen forms) and the Jeans mass drops dramatically. At this point the baryons begin to fall into the potential wells already formed by the neutrinos. The epoch of decoupling and the point where matter begins to dominate the density of the universe (instead of radiation) are indicated on the top axis. The free-streaming mass, M_{fs}, the mass within the horizon at the time when neutrinos become non-relativistic, is indicated.

becoming very worrisome. For example, observations by Juan Uson and Dave Wilkinson (1982) at Princeton had pushed the limits down to $\delta T/T_0 < 10^{-4}$ – the temperature varied by less than one part in 10 000 on an angular scale corresponding to regions containing the masses of galaxies or clusters of galaxies at $z = 1000$ – and nothing was seen. And yet fluctuations of this magnitude seem to be required if structure is to form by gravitational instability in the baryon–photon fluid only.

At this point, a number of cosmologists realized that non-interacting non-baryonic dark matter, neutrinos or the hypothetical more massive particles discussed by Gunn and collaborators, could resolve this contradiction between expectations and observations. This realization occurred to several people at about the same time – Peebles (1982) at Princeton, Nicola Vittorio and Joseph Silk (1984) at the University of California in Berkeley, Richard Bond, George Efstathiou and Alex Szalay (1983, 1984) at Stanford and Berkeley. And the solution was basically what had been outlined previously with respect to neutrino dark matter. If a non-interacting dark matter fluid exists with $\Omega_{dm} \approx 1$ then fluctuations in that fluid could begin to collapse when matter becomes the primary gravitating fluid – at $z \approx 10\,000$ rather than $z \approx 1000$ when baryons decouple from photons. This means that at $z = 1000$ the dark matter fluctuations can have grown to $\delta\rho/\rho \approx 10^{-4}$ while the photon–baryon sound wave fluctuations are stuck at 10^{-5} consistent with observations. After decoupling the baryons then quickly catch up to the dark matter, but the fluctuations in the CMB temperature are still frozen at 10^{-5}.

It is remarkable that shortly after a perceived need for dark matter on the scale of galaxies, it appeared that dark matter is also required on the scale of the Universe at large – not dark matter consisting of low-mass or dead stars, ordinary baryonic dark matter, but dark matter comprised of subatomic particles which interact with ordinary matter primarily via the force of gravity. Moreover, the primary function of this cosmological dark matter is to promote the formation of structure on small scales – so by this reasoning, it must be the same sort of dark matter that is responsible for flat rotation curves in spiral galaxies. What, then, are the astrophysical constraints upon the properties of this dark matter?

6.3 Some like it hot, most like it cold, all like it in the pot
10 billion years old

Non-baryonic dark matter certainly exists in the form of neutrinos. Cosmological neutrinos are present in the Universe with a number density comparable to that of CMB photons. Moreover, they have mass – at least 0.05 eV and possibly as much as 2 eV. Twenty years ago, before the tighter experimental constraints on the neutrino mass, it was thought that the mass could be as large as 20 or 30 eV. If this were true then it could be the case that $\Omega_\nu \approx 1$; in other words, neutrinos could be the dark matter that fills the Universe as well as the dark matter in galaxies (Bond, Efstathiou and Silk 1980; White, Frenk and Davis 1983). So the neutrino was the first plausible non-baryonic dark matter candidate. Very soon, however, around 1980, two quite serious problems with neutrinos became apparent.

First of all, there is the problem that structure on relatively small scales (galaxy scale) forms rather late in the history of the Universe. Neutrinos decouple from photons when the mean energy of particles and photons is about 2 MeV. Particles with a rest-mass of a few electron volts at most but with a kinetic energy of 2 MeV are highly relativistic: they are moving essentially at the speed of light and freely stream to the horizon. Such dark matter, that is extremely relativistic when it decouples, is called "hot dark matter".

For hot dark matter, all fluctuations on scales smaller than the free-streaming distance are washed out (see Fig. 6.2). The free-streaming distance is essentially the

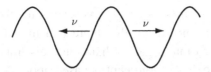

Fig. 6.2. Neutrinos, when relativistic, freely stream out of positive density fluctuations. The fluctuations are erased by the process out to a scale comparable to that
of the horizon at the epoch when the neutrinos become non-relativistic. This is a
general characteristic of hot dark matter.

scale of the horizon at the point where the particles become non-relativistic, that is,
where the average kinetic energy drops to the rest-mass energy. For hypothetical
neutrinos with an assumed rest-mass of 17.5 eV (as in Fig. 6.1) this occurs at a
redshift of about $z = 10^5$. The Universe is about 700 years old at this point and the
free-streaming distance, the horizon, would be 700 light years or about 200 parsecs.
This scale would have by now expanded to 20 Mpc and would enclose a mass of
almost $10^{16} M_\odot$. This corresponds to the peak of the curve marked M_{Jb} in Fig. 6.1.
All fluctuations in the neutrino fluid smaller than this mass scale are erased by the
free streaming of the neutrinos while relativistic. In other words, the lowest-mass
objects that could first partake in gravitational collapse, at a redshift of 20 000 when
the matter begins to dominate, are comparable to that of superclusters of galaxies
(clusters of clusters). This does not mean that lower-mass objects do not form; it is
just that high-mass objects, superclusters and clusters form first, and then galaxies
fragment out of this larger structure. It would be a hierarchy of "top-down" structure formation that was favored by Soviet cosmologists led by Zeldovich (1977).

Structure formation can actually be simulated in a computer. Around 1980 a new
science of experimental cosmology emerged, and prominent among the original
practitioners were Marc Davis, George Efstathiou, Carlos Frenk and Simon White,
as well as Anatoly Klypin and Sergei Shandarin in the Soviet Union. The cosmologist takes a representative volume, usually a cube, of the Universe, populates it with
point masses interacting only via gravity and gives the entire ensemble a uniform
expansion to simulate the Hubble flow. In order to form structure, the density distribution of particles must contain fluctuations, so the experimental cosmologists
take the distribution of fluctuations produced by inflation but with a cut-off below
some scale corresponding to the free-streaming scale of neutrinos; the density is
taken to be completely smooth below this scale.

Such simulations immediately revealed that the galaxy-mass objects form rather
late, at a redshift of $z \approx 2$, and we know that in the real Universe galaxies existed
long before this because they are directly observed at higher redshift. It is a fundamental problem with hot dark matter such as neutrinos: galaxies form much too
recently in the history of the Universe.

There is a second problem which is just as severe. In 1979 Scott Tremaine and James Gunn at Caltech pointed out that there is a limit to the number of neutrinos that can be packed into a galaxy halo, and this limit depends strongly on the mass of the neutrino. Their argument is based upon the theorem of statistical mechanics, the Liouville theorem, which is a sort of classical version of the Pauli exclusion principle. Essentially, the statement is that the density of neutrinos in phase space, a six-dimensional space consisting of the usual three spatial coordinates (x, y, z) and three momentum coordinates (p_x, p_y, p_z), can never exceed its original value (when the neutrinos decoupled from photons). This means that the maximum density of neutrinos comprising a halo depends upon the velocity spread of the neutrinos in the halo as well as their mass: the larger this velocity dispersion and the higher the neutrino mass, the higher the mass density of neutrinos in the halo. If neutrinos are to form a dynamically significant galaxy halo with a velocity spread of 100 km/s, the mass of the neutrinos must exceed 30 eV. At the time of the Tremaine and Gunn paper, this was already in excess of the experimental limit on the neutrino mass. In other words, neutrinos could not comprise the halos of low-mass galaxies.

This was fairly devastating for the proposal that ordinary neutrinos comprise the dark matter in galaxies. Indeed, the structure-formation argument seems to rule out hot dark matter altogether. The alternative is "cold dark matter" or CDM. With respect to the origin of this idea, the major players here were again those mentioned above; in particular, Peebles presented arguments for a cold pressureless medium as the driver of structure formation in 1982. The distinction between hot and cold dark matter with respect to structure formation was clearly described by George Blumenthal, Heinz Pagels and Joel Primack in 1982 and by Richard Bond and Alex Szalay in 1983. In 1984 the case for CDM, particularly with respect to galaxy formation, was codified in a very influential paper by George Blumenthal, Sandra Faber (UC, Santa Cruz), Joel Primack (Stanford) and Martin Rees (Cambridge).

Cold dark matter consists of particles that are non-relativistic when they decouple from photons. For example, a hypothetical particle with a mass (measured in terms of energy) of 100 billion electron volts (100 GeV or giga-electron volts) that decouples when the temperature is 10 GeV would be non-relativistic; such particles would be moving with a velocity far less than the speed of light. These slow-moving particles do not erase the density fluctuations on a small scale so the original spectrum of fluctuations delivered by inflation is preserved. In particular there is no lower limit on the mass of objects which can first gravitationally collapse, and that means that when a density fluctuation in the CDM fluid with some particular size scale becomes smaller than the horizon, it does not oscillate like a sound wave but continues to grow by gravitational instability.

This point requires a bit more explanation: the physical size of any density fluctuation, λ_f, is growing with the expansion of the Universe – as the square root of time in the radiation-dominated period ($\lambda_h \propto \sqrt{t}$). But the horizon itself, a causally connected region, is growing linearly with time ($l_H \approx ct$). Therefore the horizon is always catching up with fluctuations on larger and larger scales; fluctuations are said to "enter the horizon" and become causally connected, smaller ones before larger ones. Fluctuations which form galaxies and clusters all enter the horizon during the radiation-dominated period (before $z = 10^4$ in an $\Omega_0 = 1$ universe). Fluctuations in the baryon–photon fluid, after they enter the horizon, do not grow but oscillate because they are smaller than the Jeans length; they do not collapse but maintain the same amplitude ($\approx 10^{-5}$) as when they enter the horizon. But the Jeans length in *cold* dark matter is zero (zero sound speed) so, in this component, the fluctuations continue to grow. This leads to a scenario of structure formation which is more "bottom-up". Small structure forms first – objects with the mass of low-mass galaxies, some of which merge leading to larger and larger objects, a process of bottom-up hierarchical structure formation.

This point is well illustrated in Fig. 6.3 based upon calculations by Efstathiou and Bond. Here we see that in the radiation-dominated epoch, the super-horizon fluctuations in the non-interacting CDM fluid and in the photon–baryon fluid are slowly growing together (recall, that these are like separate universes in which over-densities expand more slowly than the Universe on average). When a fluctuation on some given scale (that of galaxies or clusters of galaxies) becomes smaller than the horizon, the baryon–photon fluid on that scale begins to oscillate. These oscillations are, in fact, the sound waves predicted by Jeans for fluctuations which are smaller than the Jeans scale for gravitational collapse. But the fluctuations in the non-baryonic fluid keep on growing. At decoupling of matter and radiation (when z = 1000) the photons can then freely stream to be observed 13 billion years later by us, but the baryons, no longer bound to the photons, fall into the gravitational wells of the non-interacting component. Thus the photons reflect the amplitude of the fluctuations on a small scale at $z \approx 10^4$. In order to grow to a present amplitude of $\delta\rho/\rho = 1$, i.e., to form galaxies and clusters, the original amplitude of the fluctuations need only be $\delta\rho/\rho \approx 10^{-5}$, consistent with observations.

Such a model overcomes both of the objections to hot dark matter described above. The cosmic N-body calculations demonstrate that, because small objects form first; galaxies can form early, well before a redshift of two. Moreover, because the matter is originally cold – with a very low velocity spread – there is no limit on the density of such particles in a galaxy halo. There is no problem packing cold dark matter particles into very low-mass galaxies.

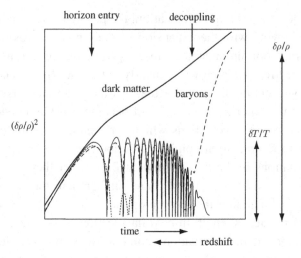

Fig. 6.3. The growth of fluctuations in a dark matter dominated universe. The amplitude of the fluctuation is plotted against the increasing age of the universe (or decreasing redshift). The solid curve shows fluctuation in the dark matter component, the dashed curve is that of the baryons, and the oscillating solid curve is the photon perturbation. The fluctuation shown here would have now expanded to 1 Mpc in the absence of gravitational collapse, and; contains a mass corresponding to that of a typical galaxy. The fluctuation becomes smaller than the horizon scale at a redshift of about 100 000, but while the dark matter fluctuation keeps growing, that in the baryon–photon fluid oscillates as a sound wave. At decoupling, when the protons and electrons combine to form (primarily) hydrogen ($z \approx 1000$), the baryons rapidly fall into the dark matter potential wells but the photons free stream to be observed by us 13 billion years later. Thus the temperature fluctuations reflect the fluctuation amplitude at the earlier epoch ($z = 100\,000$) and are much smaller than the actual density fluctuations. This figure is based upon a calculation by Efstathiou and Bond (1986).

6.4 What is the matter?

Particles that interact very weakly with baryons and photons and that decouple from the expanding photon–baryon fluid when they are non-relativistic (cold) can solve two major perceived problems: first, the observed structure on a wide range of scales can form from the gravitational instability of very small fluctuations in the density distribution beginning at decoupling ($z = 1000$). The amplitudes of these required fluctuations are consistent with those observed in the CMB. Second, these same particles can comprise the dark halos evidenced by flat rotation curves of spiral galaxies. But what are the particles? What are their properties? Are they among the known subatomic particles? If not, how might we detect them independently of their gravitational influence?

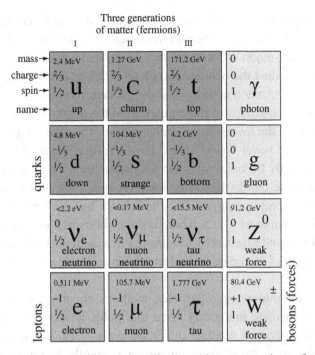

Fig. 6.4. The standard-model particles. There are three generations of quarks and leptons which all have a spin of one-half. The force-carrying gauge bosons are shown on the right. The photon carries the electromagnetic force, the gluon the strong force, and the W and Z bosons the weak force. These particles all have spin 1. The graviton, which has never been detected, is not shown and has spin 2 (from Wikipedia, author MissMJ).

In particle physics there is a standard model that has been very successful, not only in explaining many aspects of high-energy experiments, but also in predicting the existence of new particles that have subsequently been found.

The particles of the standard model are shown in Fig. 6.4. There are two general sorts of particles: fermions and bosons which are distinguished by a property called "spin". Spin is an appropriate description because it determines the angular momentum of a particle, and it is quantized – it comes in distinct lumps with magnitude $h/2\pi$, where h is Planck's constant; fermions have half integral spin (1/2) and bosons have integral spin (0, 1, 2).

Examples of fermions are electrons and baryons (protons and neutrons). But, while electrons are fundamental (not divisible), baryons are composites of fundamental particles called quarks. The electron belongs to a class of fermions called leptons. There are three kinds, or families, of leptons – electrons, muons and tau particles, but only the electron is stable (the other two decay in short order). The

electron, as we gather from its name, is charged and so interacts via the electromagnetic force. The three leptons also have their anti-particles – the anti-electron
or positron, the anti-muon and the anti-tau. Associated with each charged lepton is
a neutral particle called the neutrino, so there are three flavors of neutrinos – the
electron neutrino, the muon neutrino and the tau neutrino (and anti-neutrinos).

There are also two groups of three quarks, but the quarks which typically comprise the baryons – the lowest-mass, stable quarks – are the up quarks and down
quarks. Quarks are never found alone; the force that binds them together, the
strong force, actually becomes stronger with separation. Three such quarks make
a baryon, and these quarks have fractional charges (that is, fractional in terms
of the electron or proton charge). The "up" quark has a charge of $+2/3$ and the
"down" quark a charge of $-1/3$. Therefore a combination of two up quarks and
one down quark bound together by the strong force comprises a proton with charge
$+1$. A combination of one up quark and two down quarks is a neutron with a charge
of zero.

The bosons, the integral spin particles, mediate, or carry, the forces of nature:
gluons (spin 1) mediate the strong force that holds quarks together in baryons;
W and Z (spin 1) bosons mediate the weak force – also a "nuclear" force; the
photon (spin 1) carries the electromagnetic force; and the graviton (spin 2) carries
the gravitational force; and finally the hypothetical Higgs boson (spin 0) generates
the mass of particles.

Which of these particles has the right properties to comprise the cold dark matter? The particle should not be charged, otherwise it would be easily detectable in
electromagnetic interactions (it would not be weakly interacting). This of course
eliminates the electron and the proton (the stable configuration of quarks). The
particle should be stable – long-lived – and this would eliminate the neutron (an
unstable configuration of quarks), the tau particle, the muon, the W and Z bosons.
The particle should have non-zero mass and this would rule out the photon or the
graviton. The only particle left in the standard model zoo is the neutrino; it is electrically neutral, it has mass and has a large cosmic abundance. But for the reasons
we have discussed above, neutrinos cannot be the primary component of the putative dark matter which promotes structure formation and explains galaxy rotation
curves. There is no standard-model particle which could possibly be the cold dark
matter.

Theoretical physicists are convinced that the standard model is not the final theory. There are a number of phenomena which find no explanation in the context of
the standard model and must be added in an ad hoc manner. For example, the Higgs
mechanism, the mysterious field which gives mass to all other particles, does not
follow in any sense from the standard model. The apparent asymmetry between
matter and anti-matter is not explained by the standard model. Neutrino masses do

Table 6.1. *Standard model particles and superpartners*

SM particle	Symbol	Spin	Superparticle	Symbol	Spin
electron	e	1/2	selectron	$\tilde{e}$	0
muon	μ	1/2	smuon	$\tilde{\mu}$	0
tau	τ	1/2	stau	$\tilde{\tau}$	0
neutrino	ν	1/2	sneutrino	$\tilde{\nu}$	0
quark	q	1/2	squark	$\tilde{q}$	0
photon	γ	1	photino	$\tilde{\gamma}$	1/2
W boson	$W^\pm$	1	Wino	$\tilde{W}^\pm$	1/2
Z boson	Z	1	Zino	$\tilde{Z}$	1/2
gluon	g	1	gluino	$\tilde{g}$	3/2
graviton	G	2	gravitino	$\tilde{G}$	3/2
Higgs boson	h	0	Higgsino	$\tilde{h}$	1/2

not naturally arise in the context of the standard model. There is clearly physics, a deeper theory, beyond the standard model.

The most favored successor to the standard model is "supersymmetry", SUSY, which postulates a symmetry between fermions and bosons. That is to say, for every particle with integral spin, for every boson, there is a partner with half integral spin, a fermion; and vice versa. For example, for the photon with spin 1, there would be the photino with spin 1/2. For the electron with spin 1/2, there would be the selectron with spin 0. Every known particle would have its supersymmetric partner – a particle with spin differing by 1/2. The standard model particles along with their hypothetical superpartners are shown in Table 6.1. Such a theory would resolve the problems pointed out above, but at the expense of doubling the number of particle species. The world would be populated by twice as many types of particles which, of course, raises the possibility of dark matter candidates.

The obvious problem is that none of these hypothetical particles has yet been detected. The solution to this problem is that supersymmetry is a symmetry that is restored (or broken depending on how you look at it) at very high energy – perhaps in excess of 1000 GeV (TeV scale). Putting it another way, when the Universe was extremely young, say $t < 10^{-6}$ seconds and the energy of particles was in excess of 1000 GeV, then supersymmetric partners were present in equal abundance to the "normal" standard-model particles that inhabit the Universe at present. But, as the Universe expanded and cooled, and the energy fell below 100 GeV, almost all of the supersymmetric partners decayed leaving only the standard-model particles. Almost all – the lowest-mass supersymmetric partner has nothing to decay into and is stable.

Thus supersymmetry provides a plausible dark matter candidate, the "lightest superpartner" – the LSP. What could it be? It should be electrically neutral, so that would seem to rule out the partners of the electron or the quarks. Possibilities are the "zino" which is the fermionic partner of the Z boson that carries the weak force; the "photino", the spin 1/2 partner of the photon; and the "higgsino" the partner of the Higgs boson which generates mass. It turns out that these three particles actually "mix". The phenomenon of mixing occurs when two or more particles share certain properties such as mass, electric charge, weak charge or spin. Then if the other unshared properties are not conserved, these particles will mix with one another: an individual photino or zino or higgsino is not detected but rather realized as a quantum mechanical mixture of these particles called a neutralino. The neutralino is likely to be the lightest supersymmetric partner, stable (long-lived) and therefore a well-motivated supersymmetric dark matter particle.

We should realize, however, that supersymmetry is at this point a theory without direct experimental confirmation. Moreover, if supersymmetry is indeed the deeper theory underlying the standard model, this does not mean that the dark matter is the lightest superpartner. That particle, the neutralino for example, must have the right properties – mass and cosmic abundance – to be the dark matter. The neutralino is one of a number of hypothetical dark matter particles called, generically, WIMPs – for "weakly interacting massive particles". There are other plausible candidates that could behave as cold dark matter – notably a theoretically motivated particle with very low-mass, the axion (a non-WIMP), but the LSP is probably the leading candidate.

6.5 A new paradigm: standard CDM

The observation of mass discrepancies in spiral galaxies and clusters of galaxies circa 1980 combined with the cosmological arguments for a substantial non-baryonic component to the Universe led to the emergence of a new paradigm – the standard CDM model of the Universe. Here, the Universe was perceived to be at the critical density $\Omega_0 = 1$, consisting of 5% baryonic matter – perhaps 10% of this actually in visible form – and 95% non-baryonic dark matter. At the same time as the emergence of these cosmological arguments, the development of supersymmetry provided a plausible non-baryonic dark matter candidate in the form of the lowest-mass supersymmetric partner. The cosmologists and theoretical physicists were as strangers in the night, who found each other, discovered their shared mutual need for a major new component of the Universe, and spawned a new field, astroparticle physics – a field with a theoretical wing, calculating, for example, the locally predicted flux of halo dark matter particles, and an experimental side which goes forward with various strategies for direct detection of the dark matter particles.

Physicists were understandably excited about the possibility of non-baryonic dark matter: "It was realized in the early 1980s that Supersymmetry predicted that there should be considerably more LSP dark matter than even the matter in the stars" writes Gordon Kane in his book *Supersymmetry* (2000). This rather gives the impression that theoretical physicists predicted the existence of dark matter and that dark matter is an inevitable consequence of supersymmetry. This is inaccurate on both counts. No physicist ever told astronomers to go out and look for the gravitational signatures of dark matter. This was quite an independent development, as we have seen. Moreover, supersymmetry may be the correct theory and the LSP may exist; but that does not imply that the LSP is the dark matter. That depends entirely upon the properties of this particle, primarily its annihilation cross section, and this is not specified by the theory; it is an experimental issue.

The primordial abundance of WIMPs is set by the rate at which WIMPs and anti-WIMPs annihilate one another in the early Universe and this is determined by the annihilation cross section. The scenario is this: suppose the LSP has a mass of 50 GeV. Then when the temperature of the Universe is higher than 50 GeV, that is to say, when the photons and particles have an energy in excess of the WIMP rest-mass energy, then photons, leptons and WIMPs are continually interacting and changing from one to another. But as the Universe expands and cools, and the photon energy falls below the WIMP rest-mass energy, then the WIMPs freeze out of this soup – they became decoupled from the photons. As the Universe expands and cools further, the WIMPs begin to disappear because of self-annihilation which is determined by this cross section. It turns out that the present density of WIMPs in the Universe, Ω_w, depends only upon the annihilation cross section and not directly upon mass. The relation is

$$\Omega_w \approx \left(\frac{10^{-37} \text{cm}^2}{\sigma_a} \right) \tag{6.2}$$

where σ_a is the annihilation cross section. If the cross section is that of the weak interactions then it will be of the order of 10^{-37} cm^2, so it is possible that such particles could dominate the mass density of the Universe.

This provocative coincidence is often cited as the primary reason for supposing such particles would comprise the dark matter. But we should bear in mind that it could also easily be a factor of 100 less than the critical density. Supersymmetry does not require dark matter with a cosmologically significant density. It is, however, of great experimental importance that the hypothetical LSP will also elastically scatter nucleons – the nuclei of atoms – and that gives rise to the possibility of direct detection. In Chapter 11 I will describe the strategies and results so far for direct detection.

Suffice it to say, the cosmological arguments supported by the emergence of plausible candidates for cold dark matter particles led to the paradigm of standard CDM. This paradigm did rather quickly run into difficulties and has now been replaced (by "lambda CDM"), but it did allow for calculations relevant not only to cosmological structure formation but also to galaxies and rotation curves. Dark matter had become predictive, and in Chapter 8 I will consider the confrontation of CDM with the observations of galaxies. First, however, I return to the site of the original dark matter problem: clusters of galaxies.

7

Clusters revisited: missing mass found

7.1 The reality of the cluster discrepancy

Following the developments that I have described it was generally accepted that the discrepancy in clusters of galaxies, found by Zwicky 50 years earlier, is not an artifact of poorly understood cluster dynamics nor an aspect of bizarre new physics such as ejection of galaxies by a parent galaxy, but is actual and related to the discrepancy in galaxies revealed by rotation curves. In an influential paper in 1979, Sandra Faber (UC Santa Cruz) and Jay Gallagher (University of Illinois) reviewed the evidence for missing mass on scales ranging from galaxies to the giant clusters of galaxies. By applying the traditional virial theorem to giant clusters like that in Coma, they pointed out that the dynamical mass typically turns out to be on the order of 10^{15} $M_\odot$ while the luminosity in visible galaxies is more like 10^{13} $L_\odot$. Thus the mass-to-light ratio in clusters is of the order of 100 as originally estimated by Zwicky. If the stellar component of clusters (primarily elliptical galaxies and gas-free disk galaxies) has a mass-to-light ratio of 10, a generously large value, then at most only 10% of the mass in clusters can be in the form of normal stars (1–5% would be more likely). It had been suggested that there could perhaps be free-floating stars (not attached to galaxies) in the clusters, or very low-surface-brightness galaxies that have so far escaped direct detection. But observations of the diffuse light in the Coma cluster indicated that the total luminosity in such objects can, at most, be comparable to that in the bright galaxies.

The question, first considered by Zwicky, naturally arose: is the dark matter distributed uniformly throughout the cluster, or is it associated with the dark halos of the individual galaxies – the halos evidenced by the non-declining rotation curves of individual, non-cluster spiral galaxies? It was (and is) not possible to measure the rotation curves of cluster galaxies because most of the galaxies in rich clusters are not spiral galaxies; most are ellipticals without conspicuous disks or extended neutral hydrogen. In 1965 Herbert Rood of the University of Michigan pointed out that, based upon dynamical modeling of the mass distribution, the unseen mass seems to be distributed in the same manner as the galaxies. But there were convincing arguments given by Rood and later by Simon White (1977) that the dark mass could not be associated with individual galaxies, at least not if the total mass of a galaxy, including the dark mass, were proportional to the luminosity of the galaxy. In that case, due to gravitational interactions between galaxies, the more massive and therefore more luminous galaxies should sink to the center of the cluster. This means that the more luminous galaxies should be observed in the center of the cluster, and this is not the case, at least not in the Coma cluster. The visible galaxies are apparently swimming in a sea of dark matter.

As with individual galaxies there were speculations about the nature of this dark matter. The early suggestions, such as that of Ostriker, Peebles and Yahil (1974), centered on low-mass, low-luminosity stars – or even planetary-size objects. Another possibility, often discussed, was that of non-radiating remnants of expired stars – cold white dwarfs, neutron stars or black holes. With the increasing acceptance of the idea of a ubiquitous, universal fluid of weakly interacting elementary particles, CDM became the natural medium for the dark component of clusters. In fact, if CDM exists with the supposed properties and universal abundance, then it must constitute the dominant component of clusters as well as individual galaxies.

The advent of orbiting X-ray observatories in the 1970s opened a new window on high-energy phenomena in the Universe, and provoked a great leap forward in the understanding of clusters of galaxies. In fact, a substantial fraction of Zwicky's missing matter in clusters was actually found; it is present as hot X-ray emitting gas.

7.2 Hot gas in clusters of galaxies

In his prescient paper in 1963 (see Chapter 2), Arrigo Finzi considered the possibility that undetected mass in clusters of galaxies may be in the form of hot gas. He pointed out that the gas would have a temperature such that the thermal velocity dispersion is comparable to that of the galaxies in the cluster – typically 1000 km/s. That translates into 10 million kelvins. This hot gas should therefore be emitting X-rays by thermal Bremsstrahlung. Finzi stressed that the hot gas cannot provide the matter necessary to bind clusters of galaxies because, in that case, the density would be so high that the gas would cool in less than a million years – far short of the cosmological timescale necessary. This was all very perceptive because in the early 1970s X-ray emission from clusters of galaxies was actually discovered and soon after the source of the X-rays was identified as hot gas. And Finzi was also right about the mass of gas being insufficient to bind clusters; although, the density is sufficiently large in a number of clusters for cooling to be taking place in the central regions.

It is fortunate for life on Earth that X-rays do not penetrate the atmosphere – fortunate for life, but difficult for astronomers who would like to observe electromagnetic radiation at very short wavelengths (or high energy). In the late 1960s several groups began hoisting X-ray detectors to high altitude using balloons or rockets. The initial detectors were hardly more than sensitive Geiger counters with no image-forming capability. These provided brief fleeting views of the X-ray sky before descending and could hardly qualify as X-ray observatories. The angular resolution was on the order of degrees, so sources of X-ray emission could not be located with precision. Nonetheless, with such relatively crude instrumentation discrete sources were detected – objects within the Milky Way as well as extragalactic sources associated with active galactic nuclei and clusters of galaxies.

Two groups were prominent in this early work: a group at the Naval Research Laboratory under Herbert Friedman (a pioneer in rocket astronomy), and a group at American Science and Engineering, a research institute in Cambridge Massachusetts, under Riccardo Giacconi (for his pioneering work in X-ray astronomy Giacconi was awarded the Nobel prize in 2002). A breakthrough in this field came with the "Uhuru satellite" launched in 1970 in Kenya (on the anniversary of Kenyan Independence and consequently given the name "freedom" in Swahili). Uhuru was the first proper X-ray observatory on an orbiting platform and covered an energy range of 2 to 20 keV. The angular resolution of one-half degree was a considerable improvement over earlier detectors. Uhuru quickly discovered that clusters of galaxies were the most common extragalactic X-ray sources having huge luminosities of 10^{43} to 10^{45} ergs/s at X-ray energies.

The actual mechanism of the X-ray emission was not so immediately obvious. There were two possibilities: thermal emission, "Bremsstrahlung", from hot gas (basically due to the acceleration of electrons in encounters with other charged particles) and non-thermal radiation from relativistic electrons (so-called inverse Compton radiation whereby low-energy photons are scattered up to X-ray energies by the relativistic electrons). These two processes may be distinguished by their different continuum spectra (the distribution of X-rays by energy): the thermal radiation has an exponential spectrum (intensity $\propto \exp(-h\nu/kT)$ where ν is the frequency and T is the temperature), and the non-thermal emission has a power-law spectrum (intensity $\propto \nu^{-\alpha}$). The problem was that the initial detectors could not easily distinguish between these two possibilities. But by 1975 it was generally recognized that, over a wide range of X-ray energies, the thermal model, X-ray emission from hot ionized gas at a temperature between 10^7 and 10^8 degrees kelvin, provided a better description of the spectra. Now we know for certain that the X-ray emission is thermal radiation from hot gas which fills the potential well of the cluster.

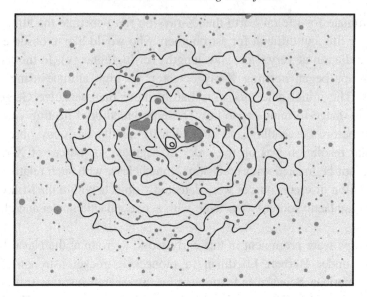

Fig. 7.1. X-ray emission from Coma cluster as observed by Jones and Forman with the Einstein satellite. The contours show equal levels of X-ray intensity and are superimposed upon an optical image of the cluster. Courtesy of Christine Jones.

X-ray astronomy and, in particular, the study of clusters of galaxies made a giant leap forward when NASA launched the Einstein Observatory in 1978. This was the first truly imaging X-ray telescope with a spatial resolution of several arc seconds. The technology of X-ray telescopes differs considerably from optical telescopes. In order to locate and image sources in the sky, it is necessary to bring the X-rays to a focus. But, unlike optical radiation, X-rays cannot be reflected by a paraboloid surface like a mirror and so are brought to a focus by grazing encounters with slightly curved surfaces. By using this technique, the Einstein X-ray telescope made actual pictures of the X-ray sky; moreover, the detectors were considerably more sensitive than previous instruments. For the first time, X-ray astronomers could determine the spatial distribution of hot gas in distant clusters of galaxies.

The surface brightness of X-ray emission as mapped by Einstein is shown for the Coma cluster in Fig. 7.1 superimposed over an optical image of the cluster. The peaks in the contour map correspond to the regions of highest gas density. The temperature of the hot gas is estimated by fitting an exponential function to the spectrum (intensity as a function of X-ray energy). Then, the X-ray astronomer determines a three-dimensional gas density distribution $\rho(r)$ by fitting a model – usually a power law extending beyond a constant density core – that reproduces the intensity map. From this model it is possible to estimate the total mass of hot gas. Moreover, given the temperature and the density distribution, it is also possible to calculate the distribution of gravitating mass.

How does this work? How do we calculate the detailed distribution of total mass in a distant cluster of galaxies using observations of the hot gas? The method relies upon the assumption of "hydrostatic equilibrium". We assume that the hot gas is at rest in the gravitational potential well of the cluster, held up against the gravitational field by ordinary gas pressure. The equation which describes this is the hydrostatic gas equation, and, assuming spherical symmetry, implies that the total gravitating mass inside radius r is given approximately by

$$M(r) \approx \frac{kT}{\mu m_p} \frac{r}{G} \frac{\Delta \rho}{\rho} \qquad (7.1)$$

where m_p is the mass of the proton and μ is the mean atomic weight; as before T is the temperature assumed here to be constant (an isothermal gas), $\Delta \rho$ is the change of the gas density over radius r and ρ is the average gas density.

Using data from Einstein, this analysis was done for a number of clusters by Christine Jones and William Forman of the Harvard-Smithsonian Center for Astrophysics in 1984, confirming earlier more tentative conclusions. They found that, first of all, there is a correlation between the temperature of the hot gas and the velocity dispersion of the galaxies in the clusters. This means that the thermal velocity of the gas particles is comparable to the random velocity of the galaxies – the two "fluids" are in thermal equilibrium with each other.

Second, in the rich clusters the total mass of the hot X-ray emitting gas ranges from 10^{13} $M_{\odot}$ to more than 10^{14} $M_{\odot}$. This typically exceeds the mass of stars in galaxies by a factor of three or four. In other words, the baryonic content of rich clusters is dominated by the hot gas and not the visible stars in the individual galaxies. The detectable baryonic matter in clusters is more accurately described as a ball of hot gas in which the subdominant galaxies are swirling about like shining pearls. Earlier work on the dynamics of clusters, going back to Zwicky, did not take this hot-gas component into account – of course, he could not because there was no way of observing the hot gas. So this is an example of some fraction of the dark matter actually being found.

However, the hot gas remains only a fraction of the dynamical mass of these systems. Applying the equation of hydrostatic equilibrium (e.g. 7.1), the X-ray astronomers found that, out to a radius of approximately one Mpc, the total dynamical mass of clusters ranges up to 10^{15} $M_{\odot}$, comparable to that implied by the analysis of the motions of the galaxies; in other words, the Newtonian dynamical mass of the clusters is still a factor of five or six times larger than the directly observable mass of hot gas. Nonetheless, the discrepancy between detectable mass and dynamical mass was reduced from the factor of 100 found by Zwicky, to a factor of about six.

Solving the hydrostatic gas equation not only tells us the total mass of the cluster, but it also permits a detailed description of the actual mass distribution; we can actually map the radial density distribution of the unseen matter. An accurate determination requires a measurement, not only of the density distribution of the gas, but also of the distribution of gas temperature (remember pressure is proportional to the product of density and temperature). The Einstein Observatory did not provide a measurement of the spectrum that was sufficiently precise to determine the temperature as a function of radial distance from the center of the cluster. So in the earlier analyses, such as that of Forman and Jones, the gas was assumed to be at the constant temperature indicated by the average X-ray spectrum over the entire cluster.

A more detailed determination of the temperature distribution only became possible in 1999 with the launch of the Chandra satellite (named for the famous astrophysicist Subrahmanyan Chandrasekhar) which combined high spatial and spectral resolution. An example of the results of such analyses is shown in Fig. 7.2 where the density (in units of the universal critical density) for 13 X-ray emitting

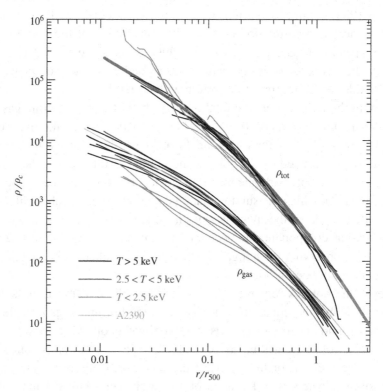

Fig. 7.2. Scaled gas- and total-density distributions for 13 relaxed clusters of galaxies. The density is in units of critical density required to close the Universe $(10^{-29} \text{ g/cm}^3)$ and the radius is in units of the radius at which the average interior density of matter in the cluster falls to 500 times this critical density. Courtesy of Alexy Vikhlinin.

clusters is plotted as a function of the scaled radius (in terms of the radius at which the density falls to 500 times the critical density). This is from work published in 2006 by Alexy Vikhlinin and collaborators.

We see that the dark-mass distribution, overall, is quite similar in form from cluster to cluster and generally more centrally concentrated than the gas-density distribution. Moreover, it appears that the unseen matter is not associated with individual galaxies but is distributed rather smoothly throughout the clusters, consistent with earlier work.

The overall fraction of baryonic mass to dark mass is weakly dependent on gas temperature, but appears to reach a maximum value of about 0.15. This has significant cosmological consequences as was noted by Simon White, Julio Navarro, August Evrard and Carlos Frenk. In 1993 they compared the total dynamical mass of the Coma cluster to the observable baryonic mass – the visible stars in galaxies and the hot X-ray emitting gas. They pointed out that within about 2 Mpc of the center, the dynamical mass is six times larger than the baryonic mass. This ratio seems to be rather typical of rich clusters. By considering the collapse of cluster-mass objects in an expanding universe consisting of dark matter and baryons, they argued that this ratio in clusters must be characteristic of the "universal ratio" of dark to baryonic mass.

Now this result is quite problematic for the standard CDM paradigm, in which it is assumed that $\Omega_0 = 1$ in dark matter plus baryonic matter. Recall that in order to produce the observed abundances of the light elements like deuterium and helium, baryonic matter must have a rather low density in terms of the closure density – $\Omega_b = 0.04$. Then, if the dark matter density is only six times larger, this means that $\Omega_{CDM} \approx 0.24$. In other words, the total density in matter, dark plus baryonic, must be of the order of $\Omega_m \approx 0.3$, only 1/3 the density required to close the Universe. This is a rather serious crack in the empirical foundation of standard CDM. Other cracks were to follow.

7.3 Gravitational lensing: a new method for probing cluster mass distribution

In 1979, Dennis Walsh, Bob Carswell and Ray Weymann, using the 2.1-meter telescope at Kitt Peak Observatory near Tucson, Arizona discovered the very first example of a gravitational lens – two images of the same quasar separated by 5.7 arc seconds. Thus began an entire new field of astronomy, a completely new tool for probing the gravitational field and, by implication, the mass distribution in distant objects such as galaxies and clusters of galaxies.

General relativity, Einstein's theory of gravity, is a truly monumental construction. It begins with a simple physical principle, the "principle of equivalence"

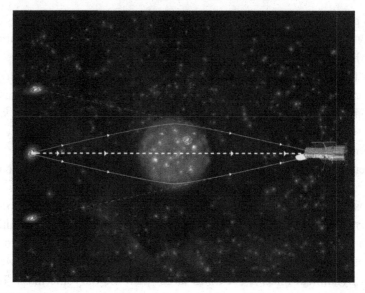

Fig. 7.3. The space telescope observing a cluster gravitational lens. The object, a background galaxy is at the right, the lens in the center, and the observer at the left. The curves show the paths of light rays and illustrate that the source is being imaged three times. In the case of perfect alignment, a circular image, an Einstein ring, would be formed. Courtesy of NASA.

(basically, there is no distinction between the effective force one feels in an accelerating frame and the force of gravity), and proceeds through a rather complicated mathematical formalism, in which the gravity field is geometrized, to a number of very definite predictions. One of these predictions is that light is deflected in a gravitational field – that a ray of light from a distant star passing near the limb of the Sun is deflected by 1.75 arc seconds. What could be more precise? And an early success of the theory was the discovery by the great British astrophysicist, Arthur Eddington, during a solar eclipse in 1919, that the images of stars near the obscured Sun were displaced by this amount.

Later Einstein considered the possibility that a star in the galaxy could produce a gravitational image of a star lying at a greater distance but far beyond: if the alignment was perfect the image would be that of a ring – an Einstein ring – about the nearer star – the lens. The geometry of such an arrangement is shown in Fig. 7.3. Zwicky, in his 1937 paper on the Coma discrepancy, pointed out that lensing by distant galaxies was in fact more probable and a possible technique for measuring the mass of the lensing galaxy. This was essentially the effect seen by Walsh and collaborators. A very distant luminous quasar (an active galactic nucleus) was almost, but not quite, directly behind an intervening galaxy which produced two images of the quasar, the double quasar. Walsh, Carswell and Weymann realized

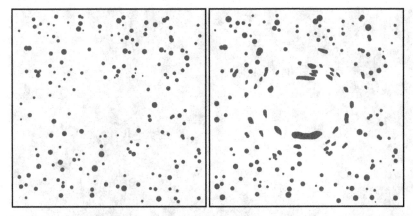

Fig. 7.4. The unlensed and lensed images of background galaxies. The figure on the left is a hypothetical distribution of background galaxies. On the right, the same background is shown after lensing by an intervening mass distribution representing a cluster of galaxies. The images of the background galaxies are stretched tangentially by the gravitational field of the cluster.

that they were observing two images of the same object due to the identical spectra, the pattern of spectral lines, in the images.

A lens arrangement in which two or more images, or an Einstein ring, are produced by the lens is called a strong gravitational lens; the act of forming multiple images is called strong gravitational lensing ("to lens" has become a new verb in the English language). But weak gravitational lensing is also possible; this occurs when the distant objects, such as galaxies, are not multiply imaged but are distorted by the intervening gravitational field as is shown in Fig. 7.4. For example, images of distant galaxies in the field of an intervening cluster of galaxies become tangentially distorted – drawn out into arc shapes as shown.

In 1986, two groups – Bev Lynds and Vahe Petrosian at Kitt Peak, Arizona and Geneviève Soucail, Bernard Fort, Yanick Mellier and Jean Picat at Toulouse – discovered that in several clusters of galaxies there appeared to be blue elongated luminous arcs in a roughly circular pattern about the center of the cluster. A number of speculations about the origin of this phenomenon immediately appeared (such as reflection by dust particles, like the ring around the Moon), but the idea that turned out to be right was due to Bodan Paczynski (1987) at Princeton. This effect is due to gravitational lensing, both strong and weak. This was established beyond doubt by the Toulouse group who obtained spectra of the faint arcs and demonstrated that they were, in fact, at much higher redshift than the clusters. The arcs are actually images of distant background galaxies that are bits of Einstein rings formed by the gravitational field of an intervening cluster of galaxies. A more recent and dramatic

Fig. 7.5. The cluster lens Abel 2218. The cluster galaxies, mostly ellipticals and gas-free disks, are the smooth randomly distributed foreground objects. The background galaxies are stretched into arcs by the intervening cluster. This object provides an example of both strong and weak lensing (NASA).

example of this phenomenon is provided by the Hubble space telescope image of the cluster, Abel 2218, shown in Fig. 7.5.

Through analysis of the pattern of tangential distortion in the images of background galaxies, it is possible to reconstruct the surface-density distribution of matter, primarily dark matter, in the lensing system. An example of such a reconstruction is shown in Fig. 7.6 for a cluster at a redshift of $z = 0.33$. This result is from the work of Henk Hoekstra, Marijn Franx, Konrad Kuijken and Gordon Squires (1998) and is based upon Hubble space telescope observations. We see the peak of the projected distribution (projected along the line-of-sight) coincides with the dominant central galaxy in this case.

The important aspect of weak gravitational lensing is that it provides a true map of the dark matter distribution. Use of the X-ray emitting gas to derive a mass distribution requires the assumption of hydrostatic equilibrium; use of gravitational lensing only requires the assumption that general relativity is the correct theory of gravity on these scales. It is not evident that the total mass and mass distributions determined by the two methods will always agree; for example, in a cluster that has just collided and merged with another cluster, the gas may well not be in equilibrium, and therefore would give a false determination of the mass. This is not so for gravitational lensing.

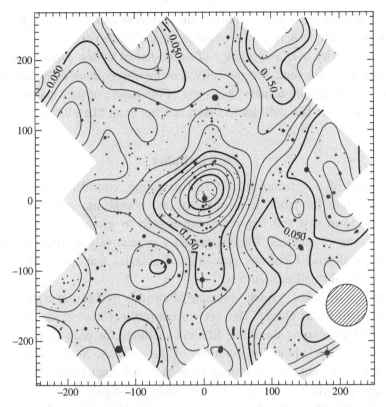

Fig. 7.6. Dark mass surface-density distribution in the cluster 1358 + 62 reconstructed from observations of the distortions introduced by weak gravitational lensing of background galaxies. The x–y units are seconds of arc and at the redshift of this cluster ($z = 0.33$); the total area covered would correspond to 1 Mpc × 1 Mpc. This map is superimposed upon an optical image of the cluster; it is evident that the overall morphology of the galaxy distribution in the cluster is also present in the dark matter distribution. From Hoekstra *et al.* (1998).

In general, for clusters with a relaxed appearance (a smooth spherically symmetric distribution of galaxies), the two methods give about the same results. In the example shown in Fig. 7.6, the lensing mass is 4.4×10^{14} M$_\odot$ and the mass derived from X-ray images and spectra is 4.2×10^{14} M$_\odot$, agreement well within the errors of the two methods and, in both cases, about seven times greater than the observed mass of X-ray emitting hot gas. There are exceptional cases in which the results are discrepant by a factor of two or three, but usually there is approximate agreement: the mass distribution probed by non-relativistic particles (individual galaxies or hot-gas particles) agrees with that probed by relativistic particles (photons). And the magnitude of the discrepancy between the detectable baryonic matter and the dynamical matter is the same in both cases.

This establishes the overall validity of the assumptions of virial and hydrostatic equilibrium in rich clusters. Some would argue that it also establishes the validity of general relativity over a vast range of astronomical scales, although actually it is a consistency argument. Very recently, lensing observations of a most striking object have dramatically revealed an aspect of the nature of dark matter itself: the dark matter behaves like collisionless particles, not at all like a diffuse gas.

7.4 The Bullet

In the context of the cosmological CDM paradigm, considered in the previous chapter, structure formation is hierarchical and bottom-up: small objects, such as galaxies, form first and larger structures, such as groups and clusters, form later by a series of mergers. There is evidence in relatively nearby clusters, such as Coma, for distinct kinematic groupings of galaxies that apparently represent recently acquired small sub-clusters.

But perhaps the most dramatic example of clusters in collision is that provided by the so-called Bullet cluster. This very recent result (Clowe *et al.*, 2006) shows the power of combining the three different methods for probing the mass distribution of clusters: optical observations of the galaxies, X-ray observations of the distribution of hot gas, and mapping of the dark matter distribution via the gravitational distortion of the images of the background galaxies.

This amazing object has been observed and analyzed by a large group of astronomers (several techniques, many astronomers): Douglas Clowe, Maruša Bradač, Anthony Gonzalez, Maxim Markevitch, Scott Randall, Christine Jones and Dennis Zaritsky. The cluster lies at a redshift of $z = 0.296$ and, optically, appears to be two galaxy concentrations – a main cluster and smaller sub-cluster separated by a distance (projected onto the sky) of 720 kpc. In X-ray observations made with the Chandra observatory it was found that there are indeed hot-gas components associated with each cluster, but these are not located at the cluster centers as defined by the most dense concentration of galaxies; they are between the two cluster centers. Moreover, the smaller of the two X-ray emitting components has a bow shape that is characteristic of a gas concentration in supersonic collision with another gaseous object. This pattern can be seen in the morphology of X-ray emitting gas shown in Fig. 7.7 (in red on the back cover of this volume). Also apparent here are the visible galaxies of the two clusters which have a larger separation than the gas concentrations.

This kind of configuration is exactly what would be expected in a high-velocity collision between two clusters. The galaxies of the two clusters, being effectively collisionless, pass directly through each other, but the gaseous components collide and decelerate. It is analogous to a collision between two balls of cotton candy

Fig. 7.7. The Bullet cluster. The red tint (rear cover) shows the X-emission from these two clusters of galaxies superimposed upon the background optical image. The gas lies between the two concentrations of galaxies and, for the smaller cluster, the morphology of the gas distribution has the shape characteristic of a bow shock. The blue tint (rear cover) is the mass distribution as determined by weak gravitational lensing and coincides with the visible galaxy concentrations (from Clowe *et al.* 2006).

with marbles embedded inside. The cotton candy would stick, but the marbles pass through. Moreover, from the shape of the bow shock, the velocity of the collision can be estimated: it is in excess of 4000 km/s, a very high-velocity collision indeed.

The distortion of the images of background galaxies tells us where the mass, the dark mass, actually lies. This is shown in blue on the figure reproduced on the back cover. Here we see that the dark mass coincides with the galaxies and not with the gas. This means that the dark matter is non-collisional like the galaxies. It is "dissipationless", not gaseous (unless it is in the form of very small dense gas clouds). This picture is completely consistent with dark matter in the form of subatomic, non-baryonic particles – like CDM.

This is an extremely important result. It demonstrates that applying these different methods to such an unusual object can tell us not only where the dark matter is located but also about the nature of the dark matter itself. This result has been heralded as a proof of the existence of dark matter. But it should be remembered that the implicit underlying assumption is that general relativity is the relevant theory of gravity on these scales. Gravitational lensing reveals the component of gravitational force perpendicular to the line-of-sight toward the lensed source. Conversion of that quantity into a mass requires a theory of gravity; so as with rotation curves the inferred existence of dark matter is not independent of the assumed law of gravity. The proof of the existence of non-baryonic dark matter can only come with its direct detection.

8

CDM confronts galaxy rotation curves

8.1 What do rotation curves require of dark matter?

The extended rotation curves of spiral galaxies are asymptotically flat. This is the essential result of three decades of 21-cm line observations carried out with radio telescopes – single-dish as well as interferometers. Every astronomer and physicist is familiar with this result and its interpretation as visible spiral galaxies being embedded in a more extensive dark halo. But what does this imply about the distribution of dark matter within spiral galaxies? What is required of dark matter in order to explain this essential observation?

In Fig. 8.1 we see again the well-known example of a flat rotation curve, that of the spiral galaxy NGC 2403. This rotation curve has been derived from 21-cm line observations made more than 20 years ago at Westerbork WSRT by Kor Begeman, then a student in Groningen, and it is a clear example of a flat rotation curve which extends well beyond the bright inner regions of the galaxy. Also shown are the combined Newtonian rotation curves of the observable baryonic components of the galaxy – the stars and the gas (assuming that the light traces the mass of the visible disk). The discrepancy between the observations and predictions is very evident in this figure.

In the context of Newtonian dynamics, dark matter must make up this difference between observations and expectations. What sort of dark matter distribution is required in this case?

Recall that the rotation curve is given by

$$V^2 = GM(r)/r. \tag{8.1}$$

Therefore if V is to be constant, this means that the mass enclosed within r, $M(r)$, has to increase with r: $M(r) \propto r$. This implies that the density of dark matter in the halo must fall as $1/r^2$ ($\rho \approx M(r)/r^3$). This is generally true in spiral galaxies. So while the light is falling off exponentially, the implied density of matter is falling

101

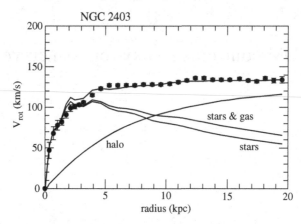

Fig. 8.1. The points show the rotation curve of the galaxy NGC 2403 as measured in the 21-cm line of neutral hydrogen. The Newtonian rotation curves of the baryonic components are also shown (stars and gas) as is the rotation curve of the halo necessary to make up the difference. The heavy solid curve is the total rotation curve resulting from the baryonic and dark components. The observations and mass decomposition are from Begeman (1987).

off much more gradually – the dark matter is becoming more and more dominant in the outer regions.

There is a historical model for a self-gravitating object in which the density falls as $1/r^2$. In 1907 the Swiss-German astrophysicist, Robert Emden (an uncle of Martin Schwarzschild) published an important book, *Gaskugeln*, gas spheres. Here Emden explored the structure of self-gravitating gaseous spheres in which there is a definite relation between the temperature and the density – a power-law relation of the form $T \propto \rho^x$. In the case where that power law is zero ($x = 0$), the temperature is constant and the resulting object is called the isothermal sphere. Emden was interested in constructing models for stars, so the isothermal sphere was not very interesting for him; stars are far from isothermal.

But there is another problem for the isothermal sphere. Because the density distribution at large radii falls as $1/r^2$ all the way to infinite radius, the mass keeps increasing as radius; that is to say, the mass is infinite. This would hardly seem appropriate for a star, but it is exactly what is needed for flat rotation curves of spiral galaxies. So the isothermal sphere, the $x = 0$ limit of Emden's gas spheres, would seem to be a perfect dynamical model for dark halos of galaxies. Of course, we would expect that even galaxies do not have infinite mass, so, for a physically realistic model, the isothermal assumption must break down; at some point the sphere must be cut off or truncated, but presumably this is well beyond the observed 21-cm line rotation curve.

But more is required of the dark halo than a density that falls like $1/r^2$. Looking again at the rotation curve for NGC 2403, we notice that the observed baryonic matter can account for most of the rotation velocity in the inner regions – there is not so much need for dark matter where the disk is bright. This has been called "the maximum-disk" (Chapter 5) because the maximum possible mass consistent with the observed rotation curve is assigned to the stellar disk. It is also possible to fit the rotation curve when the disk is sub-maximum (there is a "degeneracy" in rotation-curve fitting between halo contribution and visible disk contribution), but an argument in favor of maximum-disk is that the implied mass-to-light ratio of the stellar disk is often quite reasonable; M/L is about what we would expect for the population of stars making up the disk (at least for relatively high-surface-brightness galaxies). This requires that the dark matter distribution should not continue to increase as $1/r^2$ into the center of the galaxy, because then the dark matter would be dominant everywhere (it would also imply an infinite density at the very center of the galaxy). Therefore we would expect the halo, represented by an isothermal sphere, to have an inner core – a central region where the density does not continue to increase into the center – a region of constant density. The presence of a core is, in fact, an aspect of the mathematical model for an isothermal sphere. This density distribution is shown by the solid curve in Fig. 8.6.

The isothermal sphere with a core is the model halo that gives rise to the halo rotation curve shown in Fig. 8.1. Combined with the baryonic components, it provides a realistic representation of the observed rotation curve. But this is in no sense a prediction; it is fitting the observations after the fact. There are three free, or adjustable, parameters in such a fit – the asymptotic halo velocity, the core radius and the mass-to-light ratio of the visible disk – so it is perhaps not surprising that a reasonable match to the observations is possible.

But there is another aspect of the figure that is quite striking. In the context of the maximum-disk model the maximum rotation velocity due to the baryonic disk matter is matched by the asymptotic halo rotation curve. As first emphasized by Tjeerd van Albada and Renzo Sancisi (1986), there seems to be a "conspiracy" between the disk and the halo. As the disk rotation velocity falls, the halo rotation velocity rises just in such a way as to keep the total rotation velocity constant. This, of course, is only true in the context of the maximum-disk (it takes at least two to conspire). This aspect of observed rotation curves can be described in a different way: there is never any evidence in observed extended-rotation curves for a halo component that is in any sense distinct from the inner visible galaxy; for example, we never see a decline in rotation velocity beyond the visible disk followed by a rise connected with the unseen component. CDM must somehow confront this apparent conspiracy or the absence of a distinct halo feature.

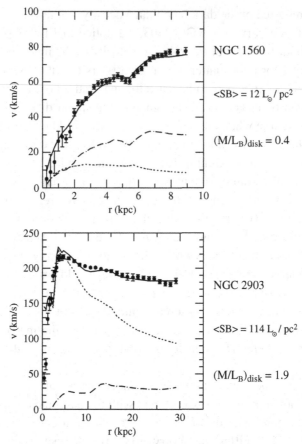

Fig. 8.2. The top panel is the 21-cm line rotation curve of the LSB galaxy NGC 1560. The points show the observed rotation curve and the dotted and dashed curves are, respectively, the Newtonian rotation curves of the stellar and gaseous disks. The mean surface brightness of the visible disk is 12 $L_\odot/\mathrm{pc}^2$. The lower panel is the rotation curve of the HSB galaxy NGC 2903 where the symbols have the same meaning. The mean surface brightness in this case is a factor of 10 larger at 114 $L_\odot/\mathrm{pc}^2$. Note that there is a large discrepancy within the optical disk for the LSB galaxy (much dark matter within the visible disk), while there is a small discrepancy in the inner regions of the HSB galaxy. The solid curves in both cases are the rotation curves predicted by modified Newtonian dynamics (Chapter 10).

With the detection and systematic observations of a large population of low-surface-brightness galaxies in the 1990s another regularity in observed rotation curves became evident: there is a systematic difference between the rotation curves of "low-surface-brightness" (LSB) galaxies and "high-surface-brightness" (HSB) galaxies. This is demonstrated in Fig. 8.2 which contrasts the observed rotation curves of two such galaxies. We see that the rotation curve of the LSB slowly rises to its asymptotic value whereas the HSB rotation curve rises rapidly and then declines to a constant value. This is a general trend in spiral galaxies that was first pointed out by Stefano Casertano and Jacqueline van Gorkom in 1991. It would be hoped that CDM could address this systematic aspect of rotation curves.

8.2 Global scaling relations

CDM must also explain the existence of well-defined "galaxy scaling relations". These are statistical correlations between the global observed properties of galaxies as defined by a large sample of galaxies. The most famous example of a scaling relation is the Tully–Fisher law (Chapter 4) that describes the correlation between the rotation velocity of spiral galaxies and their luminosity. The Tully–Fisher relation for a sample of galaxies in a nearby loose galaxy cluster, the Ursa Major cluster, is shown here in Fig. 8.3 (observed by Marc Verheijen as part of his PhD work in Groningen in 1998).

A very nice property of this sample is that these galaxies are all at about the same distance because they are members of a single cluster; there is very little relative-distance uncertainty. Moreover, the rotation velocity is measured in the 21-cm line of neutral hydrogen using high-resolution radio interferometry (Westerbork) which means that we do not have to rely upon the width of a global line profile in order to estimate the rotation velocity. The velocity plotted here is the constant rotation velocity measured at a large distance from the visible object – the "flat part" of the rotation curve. The total luminosity and the radial distribution of surface brightness

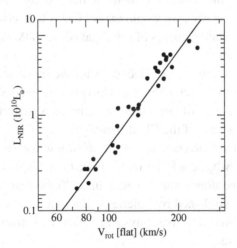

Fig. 8.3. The Tully–Fisher relationship for spiral galaxies in the Ursa Major cluster. The luminosity in the K′ band (near-infrared) in units of 10^{10} $L_\odot$ is plotted against the asymptotically flat rotation velocity in km/s (from Sanders and Verheijen, 1998). The line is not a fit to these points but shows a relationship $L \propto V^4$.

in these galaxies are measured in the near-infrared emission which is nearly proportional to the mass of the population of stars comprising the disk.

We see that this relation between luminosity and rotation velocity, of the form $L \propto V^4$, is very well defined, with very little scatter, over a factor of 100 in luminosity. This is certainly the very best correlation observed in extragalactic astronomy. There is no evidence here for an intrinsic scatter; the spread around the power-law relation is entirely due to observational error.

In the context of dark matter, the truly remarkable aspect of this relationship is that it is a correlation between the luminosity of all the stars in the galaxy, which in the near-infrared is proportional to the baryonic mass, and the rotation velocity at a large distance from the visible galaxy which is a property of the dark halo. This could be seen as another aspect of the conspiracy. How is it that a halo property correlates so well with the mass of the visible matter? CDM must somehow subsume this near-perfect correlation involving the baryonic mass and the rotation velocity established by the halo.

8.3 Structure formation in a CDM universe

CDM halos form in a cosmological context; they form via gravitational instability in the dark matter component of the expanding Universe. The halos are composed of particles which only interact by gravity; there are no gas dynamical effects (at least, not until later when the baryons cool and collapse). Therefore, it should be possible to determine the structure of realistic halos by numerically following this collapse process. In this way, we would expect CDM to become more than adjusting parameters to fit rotation curves of spiral galaxies: CDM can become predictive on the scale of galaxies.

But before considering galaxy-size objects, we must step back and reconsider the formation of structure in the Universe in general. This is important because the predicted form of large-scale structure is perhaps the most outstanding phenomenological success of the CDM paradigm.

In understanding the Universe, there is no substitute for basic astronomical data. And, in cosmology, what could be more basic than mapping the distribution of visible matter in three-dimensional space? In 1986 the first of several large-scale redshift surveys was published by Valérie de Lapparent, Margret Geller and John Huchra: the CfA (Center for Astrophysics at Harvard University) redshift survey of the nearby Universe.

This group measured the redshifts of all galaxies in a large region of the sky brighter than a certain apparent brightness or magnitude limit. Knowing the redshifts they could determine the distance (via the Hubble law) and produce a picture of the three-dimensional distribution of galaxies in the local Universe. The surprising

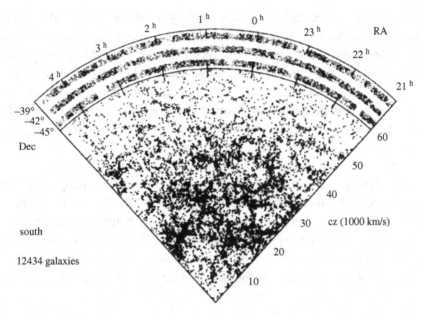

Fig. 8.4. The distribution of galaxies in a slice of the sky. The pattern is far from homogenous; obvious structure exists in the form of voids, walls and filaments. From Shectman *et al.*, 1996.

result was that the distribution of galaxies on a large scale appears to be far from homogeneous: there are large structures, walls and filaments of galaxies separated by great voids, on the scale of tens of megaparsecs – a sort of Swiss-cheese universe. This is illustrated in Fig. 8.4 which is a slice of a later and deeper survey, the Las Companos survey carried out by a large team led by Steve Shectman (1996).

The structure is very clear: the visible galaxies appear to define a porous network that has been called "the cosmic web" (a term coined by Richard Bond of the Canadian Institute for Theoretical Astrophysics). But it is most remarkable that this morphology was discovered at about the same time in N-body simulations of structure formation in the context of CDM. The very structure that observers were mapping was actually emerging in numerical simulations of the expanding Universe. This appeared to be a dramatic success for the CDM paradigm and for experimental, or numerical, cosmology.

The early growth of structure in an expanding universe can be described analytically, using only a pencil and paper, so long as density fluctuations are small ($\delta\rho/\rho \ll 1$). But when these fluctuations become large (or "non-linear") this is no longer possible; numerical methods, "experimental cosmology", becomes necessary. This has been briefly described in Chapter 6, and an early success of this technique was the elimination of neutrinos, or more generally "hot dark matter", as the dominant matter component of the Universe.

One considers an expanding volume, usually a cube with one edge having a size of 50 to 100 Mpc at the present epoch. This cube is populated with a large number of point masses interacting only via gravity. This would be a dissipationless fluid; that is to say, there are no gas dynamical effects but only Newtonian gravity which, of course, would seem entirely appropriate to the perceived nature of dark matter. It is the form of the initial density fluctuations which makes the simulation appropriate to *cold* dark matter: the fluctuations extend down to the smallest scale that can be resolved in the calculation in contrast to hot dark matter simulations where, below a certain length scale, the density of the fluid is smooth.

As the cube expands with the universe the experimental cosmologist can follow the development of structure from the linear regime where the density fluctuations are small to the point where virialized gravitationally bound objects corresponding to clusters or galaxies appear. One example of such a calculation is shown in Fig. 8.5 from a standard CDM simulation by Rien van der Weygaert. There is indeed a striking resemblance to the actual large-scale structure observed in the Universe.

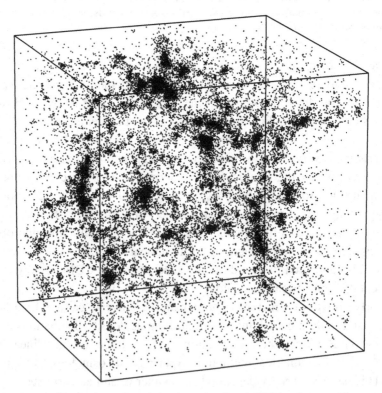

Fig. 8.5. The distribution of particles in a simulated CDM universe. (Courtesy of Rien van de Weygaert.)

Not only is the appearance similar to the actual observed structure, but various quantitative descriptions of the simulated structure also reproduce those observed. For example, in the real and simulated universes, we can measure the average amplitude of density fluctuations over various scales; i.e., the average value of $\delta\rho/\rho$ on a scale of 10 Mpc, 20 Mpc, etc. For such a quantitative measure of the distribution of fluctuations there is quite a remarkable agreement between the real Universe and the CDM simulations beginning with very small fluctuations at decoupling, consistent with the constraints imposed by the microwave background. All in all, this is very encouraging for the CDM paradigm.

8.4 The mass distribution in CDM dark halos

If the experimental cosmologist wants to determine the structure of a 10^{12} M$_\odot$ galaxy halo that develops in cosmological simulations, then that halo should consist of at least several thousand particles. This was not possible in the early simulations. But by 1990 the speed and memory of computers had increased to the point that cosmic N-body calculations could accurately follow several hundred thousand particles in an expanding cube representing a piece of the Universe (recall Moore's law). This meant that the detailed structure of relatively low-mass bound objects (i.e., galaxies) could be resolved.

So it became possible for the first time to describe in detail the galaxy halos that form by dissipationless gravitational collapse in an expanding CDM-dominated universe. Very prominent in this work were Julio Navarro (Arizona), Carlos Frenk (Durham, UK) and Simon White (Max Planck Institute for Astrophysics near Munich) who published their results in a series of papers from 1995 to 1997.

Their major discovery was that halos ranging in mass from that of small galaxies to clusters of galaxies all seem to have a characteristic density distribution – a "universal density law" – but this is not like that of an isothermal sphere. The objects that form from cold dark matter in an expanding universe do not appear to have a constant density core at the center, but rather a power-law "cusp": $\rho \propto 1/r$. The density of dark matter apparently goes right on increasing into the center. Moreover, beyond a certain critical distance from the center, r_s, this power law changes gradually from $1/r$ to $1/r^3$, so there is a range in radius where the CDM halo mimics an isothermal sphere.

This density distribution is shown by the dashed curve in Fig. 8.6, which is that used to fit the rotation-curve fit of NGC 2403 using this form for the halo (Fig. 8.8). The solid curve is the density distribution in the corresponding isothermal halo. We can see that near 10 kpc ($\approx r_s$) the density law is quite similar to that of the isothermal sphere, $\rho \propto 1/r^2$. At smaller radii, the density keeps rising into the center ($1/r$), whereas at large radii the density falls off more rapidly ($1/r^3$). This form

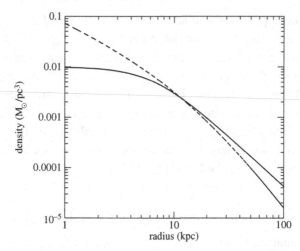

Fig. 8.6. The solid curve shows the density (in units of $M_\odot/\mathrm{pc}^3$) plotted against distance from the center of a halo modeled by an isothermal sphere. The dashed curve is the same for the NFW halo. For the isothermal sphere the core radius is 6.5 kpc and the asymptotic velocity (the flat rotation velocity at large distance) is 135 km/s. For the NFW sphere, the rotation velocity at the point, R_{200}, where the average enclosed density falls to 200 times the critical density (to close the universe) is 126 km/s, and the concentration parameter is six; the concentration parameter is defined as R_{200}/r_s where r_s is the break radius (where the density law slowly shifts from $1/r$ to $1/r^3$). Both halo models are applied to the rotation-curve fits for NGC 2403.

for the halo has become known as the NFW halo (Navarro–Frenk–White), and is now a standard feature of the CDM paradigm: it describes the predicted form of the CDM density distribution in halos of all masses, and, consequently, should be used to model the contribution of the halo to the rotation curve of galaxies.

Fig. 8.7 illustrates the rotation laws resulting from these two halo models, the isothermal sphere and the NFW sphere; i.e., these rotation curves correspond to the density laws shown in Fig. 8.6. There are notable differences in the form of these rotation laws: in the inner region the rotational velocity rises more quickly with distance in the NFW model because of the central cusp in the density distribution. In the outer regions the NFW rotation curve falls below that of the isothermal sphere because of the more rapid decline in density.

How does the NFW rotation curve perform when confronted with actual data? We see an example in Fig. 8.8, again for the spiral galaxy NGC 2403. Combining this halo contribution with that of the observed baryonic components, stars and gas, results in the total rotation curve shown here by the solid curve; this provides a reasonable match to the observed rotation curve shown by the points. So the NFW halo appears to be at least consistent with the observed rotation curves of spiral galaxies. But does this prove that it is right?

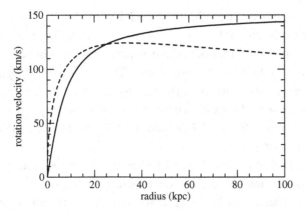

Fig. 8.7. The rotation velocity (km/s) as a function of distance from the center (kpc) resulting from the two halo models shown in Fig. 8.6. The solid curve is that of the isothermal sphere and the dashed curve is for the NFW sphere.

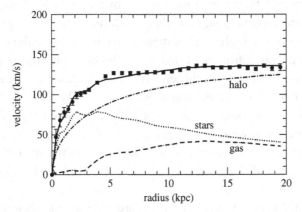

Fig. 8.8. The points show the rotation curve of the galaxy NGC 2403 as measured in the 21-cm line of neutral hydrogen. The Newtonian rotation curves of the baryonic components are also shown (stars and gas) as is the rotation curve of the NFW halo necessary to make up the difference.

As for the isothermal model, the rotation curve shown here is not a prediction; it is an exercise in curve fitting by adjusting parameters. As for the isothermal sphere there are three free parameters: a mass-to-light ratio for the disk and two parameters for the halo – a maximum halo rotation velocity which sets the mass scale of the halo, and the degree to which that mass is concentrated toward the center; a concentration parameter. This is usually described as the radius where the density falls to 200 times the critical density of the universe, R_{200}, divided by the radius where the power law changes from 1 to 3, $C = R_{200}/r_s$. Given the rather simple generic form for the observed galaxy rotation curves, it is not surprising that adjustment of these parameters for the mass model can provide reasonable fits.

This in no sense proves the validity of the NFW model. Looking back at Fig. 8.1 we see that the isothermal sphere does at least as well in fitting this particular rotation curve, and there are arguments that it performs even better, particularly for galaxies with a very low surface brightness. The low-surface-brightness galaxies appear to be completely dominated by dark matter, even in the inner regions, and usually have an observed rotation curve that is rather slowly rising from the center. This is problematic for the NFW halo model with its central density cusp and rapidly rising rotation curve.

But in fact, CDM in a cosmological setting is somewhat more predictive about the form of galaxy halos and their contribution to the observed rotation curve in galaxies; it actually can be more than an exercise in unrestricted parameter fitting. That is because such halos, formed by gravitational collapse in an expanding universe, exhibit a correlation between the degree of central concentration and the maximum rotation velocity (or characteristic mass) of the halo. This relationship emerges because, in the context of hierarchical structure formation, small objects are formed first, at higher redshifts, when the characteristic density of the universe is higher. This means that low-mass galaxies should have a higher central concentration of mass – a higher concentration parameter – than higher-mass galaxies or clusters of galaxies. Such a correlation means, in effect, that one of the free parameters of the NFW halo fit to the rotation curve vanishes – it becomes a two-, not a three-parameter fit. There is somewhat less wiggle room when there are only two free parameters in curve fitting. In fact, the concentration parameter of $C = 9$ used in the fit to model the rotation curve of NGC 2403, shown above, is about right for this velocity (or mass) scale.

The problem for CDM arises in low-mass, low-surface-brightness galaxies. This is a point made repeatedly by the radio astronomer, Stacy McGaugh of the University of Maryland and his colleagues. Fig. 8.9 shows the rotation curve of one such galaxy we have encountered before, NGC 1560. This is a dark matter dominated small galaxy observed in the 21-cm line of neutral hydrogen by Adrick Broeils. The low rotation velocity implies that the concentration parameter should be of the order of 10. With this high concentration the NFW fit to the rotation curve is shown in Fig. 8.9. It is not an excellent fit to the observations, and this can only be achieved by reducing the mass of the stellar disk to zero; the stellar component of this galaxy must have a negligible mass.

Do such rotation curves falsify CDM? There are arguments that it does not because only taking the halo implied by the dissipationless collapse of dark matter particles ignores much of the physics of galaxy formation. There are baryons in the mix as well – gas and stars. The gas collides, cools and falls to the bottom of the halo potential well and forms stars. Some of these stars are massive and become

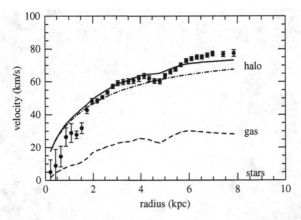

Fig. 8.9. The points show the observed rotation curve of the dwarf low-surface-brightness galaxy NGC 1560 as measured in the 21-cm line of neutral hydrogen. The Newtonian rotation curve of the gaseous component is also shown (stars and gas) as is the rotation curve of the NFW halo with the concentration implied by the cosmological-collapse models for halos. Here the visible stellar disk is assumed to have no mass.

supernovae at the end of their short lives (a supernova is the violent explosion of a star). These supernovae can blow out the remaining gas, but because the dark matter interacts gravitationally with the gas, this process may also rearrange the dark matter distribution. It is hoped that such astrophysical processes (often called "gastrophysical" processes) can rescue the CDM paradigm with respect to galaxy phenomenology.

Apart from the structure of individual halos, another conspicuous aspect of CDM halos has emerged from high-resolution N-body calculations: the presence of substructure in galaxy-scale halos.

8.5 Substructure in CDM halos

By the beginning of the new millennium computational power and techniques had improved to the point that the structure of individual galaxy-scale halos could be resolved in detail. In these higher-resolution simulations a quite new and unanticipated aspect of CDM halos appeared. This is illustrated in Fig. 8.10, which is from a modern numerical simulation by the group of Volker Springel at the Max Planck Institute for Astrophysics near Munich.

Keep in mind that this figure has nothing to do with the visible appearance of galaxies; in spite of its appearance it shows the distribution of dark matter particles. At first glance, one might think that this is a simulation of a cluster of galaxies with a dominant massive galaxy near the center. But it is a single galaxy-scale object with a mass comparable to that of the Milky Way. It is, in fact, a CDM snapshot of

Fig. 8.10. The halo which forms in an N-body simulation of a CDM universe. The central object has a mass comparable to that of the Milky Way and it is seen to be surrounded by numerous companions – sub-halos which have failed to merge with the more general structure. There are far more such objects than observed dwarf satellites of the Galaxy. Courtesy of Volker Springel.

the Milky Way halo and shows the presence of many sub-halos immersed within the general structure. That single galaxy halos contain such a wealth of substructure was first recognized in 1999, primarily by the groups of Ben Moore in Durham and Anatoly Klypin at New Mexico State University. This appears to be an unavoidable consequence of CDM.

Why should galaxy halos be so "grainy"? We have seen that in a CDM universe, the formation of halos is hierarchical: lower-mass halos form first and then higher-mass halos build up through mergers of these initial objects. Even the low-mass halos have the characteristic NFW profile. The presence of a density cusp in these objects makes them very robust; they are not likely to be torn apart in the tidal field of a more massive galactic halo. So the merging process is incomplete; much structure survives to the present day. This appears to be an inevitable consequence of CDM structure formation, and, one might suppose, leads to the possibility of falsification of the theory: if substructure is not observed, then standard CDM is wrong.

The Milky Way Galaxy does possess a number of low-mass dwarf satellite galaxies – the so-called dwarf spheroidals (more on these later) – which might be identified with such sub-halos. However, there are only about a dozen of them and the prediction is that we should see probably several hundred. Is this a falsification?

Actually, no. An obvious aspect of dark matter is that you cannot see it, or, at least, the only way that you can see it is if the dark matter potential wells contain baryons – gas which forms stars and lights up the bottom of the well. Fig. 8.10 is not a picture of the distribution of visible matter, but that of dark matter; we can only observe this structure if it contains gas and stars. We know that the principal object – the Milky Way – contains baryons and is visible, but that is not necessarily true for all the small companions. If a galaxy halo is small enough, the initial stars which form will lead to supernovae that blow away the rest of the gas, and no more stars will form. When the first generation of stars fades away, we are left with a dark galaxy.

This, at least, is one explanation of why we do not see this vast number of sub-halos (another explanation, of course, is that they are just not there). The prediction, though, is actually testable. Gravitational lensing is sensitive to mass concentrations along the line-of-sight to some source, not just to those concentrations which are visible. Several efforts are now underway looking for the predicted dark substructure via gravitational lensing in the halos of distant galaxies.

8.6 The Tully–Fisher law

As seen in Fig. 8.3, the luminosity of spiral galaxies in general is very precisely correlated with the rotation velocity at large distances from the galaxy; this relationship has the form $L \propto V^4$, at least whenever the luminosity is measured in the near-infrared light (the near-infrared luminosity is apparently tightly correlated with the stellar mass).

In the context of CDM, such a correlation must emerge from the process of galaxy formation. Somehow, the halos which form via gravitational collapse in an expanding universe and the small fraction of included baryons that contracts to form the visible galaxy must conspire to exhibit this extremely precise correlation. How is this presumed to happen?

First of all, we can ask if there is a relationship between the mass of cosmological CDM dark halos and their velocity dispersions (recall that the halos are kinematically "hot" and held up against gravity by the random motions of the dark matter particles). It turns out that there is such a relationship. This is because in CDM structure formation galaxy-scale objects (or the smaller clumps that merge and form galaxies) all form at about the same redshift and reflect the density of the universe at this time of formation. In other words, galaxy halos all have about the same density within a radius appropriate to a self-gravitating, stable "virialized" object, R_V. This means that for all galaxy halos the mass is proportional to the critical radius cubed, $M \propto R_V^3$ – some halos are more massive just because they are larger. If we combine this with the virial theorem, $V^2 = GM/R_V$, and use a bit of algebra to eliminate R_V, we find that $M \propto V^3$.

The results of simulations of CDM halos of different mass formed in an expanding universe do indeed show such a correlation, albeit with considerable scatter. Matthias Steinmetz and Julio Navarro argued in several papers (1999–2001) that this forms the basis of the Tully–Fisher law. But it is not the observed Tully–Fisher law. How do we go from this theoretical relationship between dark halo mass and velocity dispersion to a correlation between the circular velocity of the halo and the mass of baryonic material down in the center of the halo? It is necessary first to assume that the baryonic mass of a halo is some fixed fraction, f, of the dark mass (certainly in the CDM paradigm there is a fixed universal cosmological ratio of baryon to dark matter density). In the standard picture of galaxy formation this small baryonic fraction cools and collapses in the gravitational field of the dark halo. This collapse stops when a small residual angular momentum is sufficient to counteract gravity and a disk is formed (the infalling gas hits a "centrifugal barrier"). So we must also assume that the rotation velocity of the baryonic disk is proportional to the halo velocity dispersion. With two such assumptions we may derive a Tully–Fisher relation, but it would be of the form $L \propto V_{rot}^3$ with considerable scatter.

Now there is some controversy about the true exponent of the Tully–Fisher law (α where $L \propto V^\alpha$), but it certainly appears that when one measures the luminosity in the near-infrared emission (the "true" luminosity) and takes the asymptotic flat rotation velocity (as in Fig. 8.3), the exponent is 4 and not 3. This is at odds with the naive prediction of CDM.

Does this falsify the CDM paradigm? Again, CDM proponents would argue that it does not – and the argument falls back on the poorly understood aspects of galaxy formation. Gas dissipates (loses energy) and falls into the potential well of the dark halo. By this process it alters that potential well – meaning the distribution of dark matter. Moreover, stars form out of the gas, some of which become supernovae and blow away much of the remaining gas. The fraction of gas blown away depends upon the mass of the halo, so smaller mass galaxies lose more of their baryonic content (f is not constant but decreases with galaxy mass). Such processes presumably steepen the Tully–Fisher law to be consistent with what is observed and also, somehow, reduce the scatter. Perhaps so, but this is, at present, only a hope.

8.7 Can CDM be falsified by galaxy phenomenology?

It is impressive that the CDM paradigm is able to reproduce the observed appearance of the large-scale distribution of visible matter in the Universe, assuming that the visible matter is distributed in the same way on a large scale as the dark matter. CDM simulations not only reproduce the appearance of large-scale structure but also predict the observed quantitative measures of clustering, such as the present magnitude of fluctuations on differing scales. But at the same time, CDM is much less predictive with respect to galaxy phenomenology, and that is worrisome because, after all, the primary direct observational evidence for dark matter arises from the observations of galaxy rotation curves.

Usually in science, consistency is the best one can hope for; it is generally impossible to prove the correctness of a particular model, such as the NFW dark matter density distribution, which seems to be such a fundamental and inevitable consequence of CDM structure formation. But it is often possible to disprove a particular model – to *falsify* a theory. The possibility of falsification is a highly desirable aspect of any scientific theory; it is the motor of progress.

One might hope that this possibility would exist for CDM on the scale of galaxies where observations are actually quite precise. But we see that there are problems with CDM on this scale: it fails to predict the form of the rotation curves of low-mass galaxies; it leaves unexplained the systematic differences between the rotation curves of LSB and HSB galaxies; it predicts the existence of many unseen small companion satellite galaxies; it fails to produce the observed form of the Tully–Fisher law. The argument is made that these inadequacies are due to the fact that the physics of galaxy formation is not well understood; that effects other than pure dissipationless collapse become important on this scale; that when these effects are understood, the problems listed here will be surmounted.

The reliance on "baryonic physics" to bring the expectations of CDM into agreement with observations has led to the industry of "semi-analytic galaxy-formation" modeling (an excellent review of this technique is given by Carlton Baugh in 2006). Here, the poorly understood aspects of the dissipational component of galaxies – effects such as gas cooling, star formation and supernovae hydrodynamics – are modeled by simple analytic relations characterized by a number of adjustable parameters. One such parameter is that describing supernovae "feedback". In a supernovae explosion a large fraction of a star's gravitational energy is released as radiation and outward motion of the outer layers of the star. An important consequence is that the surrounding cold gas is heated and blown away; some fraction of the supernovae energy is put into the bulk motion of the gas leading to its possible expulsion from the parent galaxy. The energy fraction going into bulk motion of the surrounding gas, i.e., the feedback, is an important parameter in semi-analytic models. Other parameters are the timescale for star formation as well as exponents characterizing the initial distribution of newly formed stars by mass and the assumed power-law relation between star formation rate and the halo rotation velocity. In semi-analytic models I count as many as eight such free parameters, although some are said to be correlated by more detailed simulations involving gas dynamics or tightly constrained by observations. If an aspect of the observations is not explained – such as the distribution of galaxies by luminosity – then new physical effects are added – such as feedback due to active galactic nuclei (blowout of gas by active black holes at the galactic centers). This procedure is continued until the observations are matched. When the observations are matched, no new physical effects are added and the model is claimed to be successful.

Semi-analytic modeling has been applied in order to steepen the exponent of the Tully–Fisher relation from the predicted value of three for pure CDM halos, to the observed value of four. This is done by altering the criteria for star formation and adjusting the feedback parameter. The result is then claimed as a "predictive" success for CDM. It is true that such techniques have led to some understanding of what is required to bring the CDM scenario of galaxy formation into accordance with actual general trends in galaxy phenomenology. But more critical cosmologists see this as an exercise in parameter tuning in order to accommodate any observational result. In any case, it does appear to be an example of what Thomas Kuhn (1962) has called "normal science as puzzle solving". That is not meant in a disparaging sense because normal science is mostly what scientists do (and a number of practitioners of semi-analytic methods are very skilled at puzzle solving). But, in my opinion, the semi-analytic methods are so flexible that it is impossible for irreconcilable anomalies to be identified; such methods can always accommodate the paradigm to reality.

At present it is necessary to make a leap of faith that when the effects of baryonic physics are understood, the expectations of CDM will match the observations. This means the CDM paradigm is immune from falsification on the scale of galaxies, at least, at the present time – that the entire phenomenology of the mass discrepancy in galaxies and all of its regularities are not questions to be currently addressed by CDM. Putting it another way, the theory of structure formation in the context of CDM is presently incomplete, and this incompleteness means that CDM remains hypothetical – albeit a hypothesis that does meet most of the challenges presented by the observed large-scale structure.

9

The new cosmology: introducing
dark energy

9.1 The accelerated expansion of the Universe

In November of 1054, Chinese court astronomer, Yang Wei-te, reported to the Emperor on the appearance of a new star in the Hyades star cluster. He was, no doubt, apprehensive because unpredicted celestial events were a considerable occupational hazard for astronomers of that time and place and could lead to an abrupt and permanent termination of all contracts. What the Chinese astronomers had seen was a supernova within the Milky Way Galaxy; this particular example produced one of the most astrophysically interesting objects in the Galaxy – the Crab nebula and its embedded pulsar.

Supernovae are among the most dramatic and violent events in the Universe. A single star explodes and, for a period of time, outshines an entire galaxy. In recorded human history there are several such occurrences in the nearby Milky Way, Tycho's (1572) and Kepler's (1604) being examples observed and recorded by Europeans after the emergence of that continent from the Dark Ages.

From an observational point of view, there are clearly two types of supernovae – creatively called type I and type II. In both types, a star suddenly brightens by many orders of magnitude and then fades over several weeks. The two types are clearly distinguishable, not only by their light curves but also by their spectra: type I have none of the characteristic lines of hydrogen that are evidenced by type II. The second type, for our purposes less interesting, is thought to result from a young massive star that has exhausted its usual nuclear fuel (hydrogen) and then collapses and detonates in a spectacular nuclear explosion converting its carbon or oxygen core into iron. This process essentially destroys the star, apart from a neutron star remnant. Even though the type II supernovae can outshine a galaxy briefly, the events themselves are not so useful as a cosmological tool because the peak luminosity can vary by a factor of two – it is not a "standard candle" and therefore not useful for measuring distance.

Type I supernovae are are also exploding stars but the mechanism is different. When a normal star, like the Sun, exhausts its hydrogen, it can, after a relative brief end-of-life adventure as a red giant, quietly blow off its outer layers in a slow wind and become a white dwarf – a compact star with an extremely high density, no longer producing energy by nuclear fusion. Such an object is supported against gravity not by normal thermal pressure but by the degenerate pressure of electrons – Professor Fermi is holding up these stars. Basically, this works because electrons are fermions, their density in phase space – a six-dimensional position–momentum space – is limited to two per cell with a volume of h^3 (h is Planck's constant described in the Appendix). When the normal space density of fermions is increased beyond a critical limit, then the range in momentum, or velocity, must increase to keep the phase space density below this limit. It is this velocity due to the dense packing of the electrons and not the thermal velocity that provides the pressure support in white dwarfs. But, as discovered by Subrahmanyan Chandrasekhar in 1930, this packing pressure can only support stars below a critical mass – about one and one-half the mass of the Sun. White dwarfs cannot exist with a mass greater than the Chandrasekhar limit of 1.4 $M_\odot$.

Occasionally, a white dwarf is found in a close gravitationally bound orbit with a normal star. As the binary companion exhausts its hydrogen fuel it also becomes a red giant and dumps material from its expanding outer layers onto the white dwarf. This causes the mass of the white dwarf to grow, and it can happen that, at some point, the white dwarf mass will exceed the Chandrasekhar limit. The result is collapse and explosion – a supernova type I.

The light curves of several type I supernovae are shown in Fig. 9.1. For some reason, not entirely understood but perhaps related to the fact that a star of a standard mass is exploding, the peak luminosities of type I supernovae are comparable (top panel) – comparable but not identical: the peak luminosity is lower in supernovae with a shorter timescale for decay of the burst of luminosity. Fortunately this "stretching effect" is a well-defined correlation and can be used to "correct" the light curves to the same standard. Applying this correction, we then find the peak luminosities to be the same within about 20%; as we see in the lower panel of Fig. 9.1. With this correction type I supernovae become useful "standard candles". Moreover, because their luminosity is that of a galaxy (10^{10} $L_\odot$) and the decay time is of the order of one month, these events may be detected at cosmological distances. This makes them excellent probes of the underlying cosmology.

We all know that the apparent brightness of an object decreases as the inverse square of the distance. But this is only really true in flat space – characterized by a Euclidean geometry. Lesson one from general relativity is that space is not flat – it is curved by the matter and energy content of the Universe. So the luminosity of distant sources in the Universe will not decrease precisely as $1/R^2$; the way in

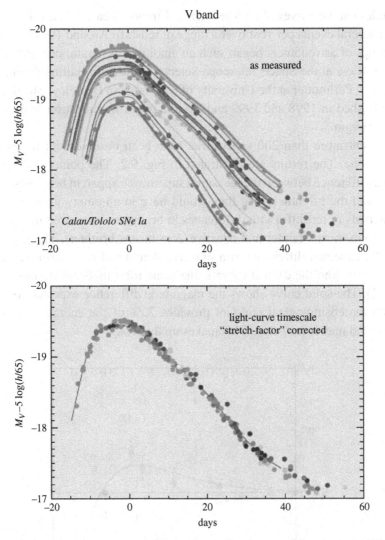

Fig. 9.1. The top figure shows the light curve of various type I supernovae (absolute brightness as a function of time). On the lower figure, the curves are all plotted corrected by the relation between peak brightness and decay time. This illustrates that these events are extremely good standard candles (from Perlmutter 2003).

which it deviates depends upon the geometry (and thus the matter content) and the expansion history of the Universe. So in principle, if we can measure the apparent brightness of many type I supernovae in distant galaxies, we can determine which Friedmann model is appropriate for our Universe.

In any single galaxy supernovae are rare events (on an individual human timescale in any case) with one or two such explosions per century. But if thousands of galaxies can be surveyed on a regular and frequent basis, then it is possible to observe several events per year over a range of redshift. Around 1994 two independent groups of astronomers began such an ambitious program: one group was led by Adam Riess at the Space Telescope Science Institute in Baltimore and the second by Saul Perlmutter at the University of California in Berkeley. The first results were published in 1998 and 1999 and led to a major modification of the standard CDM paradigm.

At present more than 200 supernovae have been observed out to redshifts in excess of one. The results are illustrated in Fig. 9.2. The points with error bars show the difference between the measured supernovae apparent brightness (in magnitudes) and the brightness that they would have in an empty universe averaged over intervals in redshift (with differences in brightness measured in magnitudes; positive Δm means fainter and negative Δm means brighter). The dotted curve shows the expected difference in a universe dominated by a cosmological constant ($\Lambda = 1$) and the dashed curve is the same for a matter-dominated universe ($\Omega_m = 1$). The solid curve shows the magnitude difference expected for a model in which the cosmological constant provides 70% of the energy density of the Universe and matter (mostly CDM) makes up the rest.

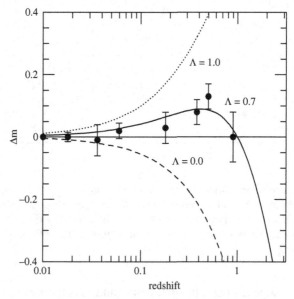

Fig. 9.2. The points are the difference between the average supernova brightness (averaged in redshift bins) and the brightness that they would have in an empty (coasting) universe. The dotted curve is the difference expected if the universe were dominated by a cosmological constant; the dashed curve is the same for a universe dominated by non-relativistic matter; and the solid curve is the magnitude difference expected in the "concordance model". In all cases the universe has the critical density ($\Omega_{tot} = 1$). The data is from Tonry *et al.* (2003).

The essential result is that type 1 supernovae are about 60% fainter than would be expected in a standard CDM universe (that is, a matter-dominated universe with CDM comprising 95% of the mass, the rest being in baryons). This appears to definitely rule out the standard CDM paradigm. But the precision of the supernova test provides an additional result that is even more astounding: the supernovae are about 20% fainter than they would be in an empty $\Omega_0 = 0$ coasting universe (a universe with no deceleration due to gravity). This means that the expansion of the Universe is actually accelerating!

In the context of general relativity, this is only possible in a universe dominated by a cosmological term in Einstein's equations – the old cosmological constant has re-emerged. The cosmological constant introduces an effective universal repulsive force in the Universe; when it dominates it leads to the accelerated expansion that we apparently observe in the supernovae statistics. The cosmological term can also be considered to be a very special kind of homogeneous fluid with negative pressure, a fluid that does not dilute as the Universe expands but maintains a constant density. Then Ω_Λ would be the contribution of the dark fluid to the present energy density of the Universe. The supernovae data are consistent with $\Omega_\Lambda \approx 0.7$ with the remainder of the $\Omega_0 = 1$ Universe being in non-relativistic matter, mostly CDM. So due to these dramatic observations, the perceived fraction of the Universe consisting of cold dark matter dropped from about 95% to 25%.

It is difficult to overstate the importance of this result. There were already problems with standard CDM – primarily the fact that in clusters of galaxies, which should fairly sample the matter content of the Universe, there appeared to be only five or six times more dark matter than detectable baryons, mostly as X-ray emitting hot gas (see Chapter 7). We know from the theory of primordial nucleosynthesis combined with observations of the abundances of the light elements that baryons can only be about 5% of the total mass density of the Universe; if there is only six or seven times more dark matter this makes the contribution of the dark matter insufficient for standard CDM – insufficient to close the Universe – in which case it should contribute 95%. On the basis of this and other arguments, primarily the tension between the measured Hubble parameter and independent estimates of the age of the Universe, Jerry Ostriker and the physicist Paul Steinhardt had previously suggested (in 1995) a "concordance model" of the Universe where a cosmological constant provided the major contribution to the expansion. But the supernovae observations provided direct evidence, through the apparent acceleration of the Hubble expansion, that the Universe is in fact currently dominated by this negative-pressure fluid – the "dark energy". Although the statistical significance of the supernova result is not high, when combined with the evidence summarized by Ostriker and Steinhardt it does point clearly to a universe with a composition that was barely imaginable a decade earlier – a universe consisting of

5% of the familiar baryonic matter comprising everything we see, 25% cold dark matter made up of undetected hypothetical particles and 70% of the even more mysterious dark energy.

9.2 COBE finds the primordial fluctuations

In 1991 NASA launched the "cosmic background explorer" satellite, COBE, into orbit and observations of the CMB emerged as the most powerful cosmological probe, not only of the early Universe but also of its evolution, structure and present matter content. The COBE satellite carried several instruments: one was a far-infrared spectrometer that could measure the absolute flux of background radiation at various wavelengths from 140 to several thousand micrometers. A second was a differential microwave radiometer that could measure the difference between the intensity of the microwave radiation in different parts of the sky separated by more than about seven degrees. Because these measurements were made in space, they were free from the contaminating radiation from the Earth and its atmosphere.

One of the initial results from the absolute spectrometer was that the spectrum of the CMB is quite precisely that of a perfect black body having a temperature of 2.73 degrees; the spectrum agreed with the theoretical Planck curve to within 0.03%. Models of the CMB based upon thermal re-radiation of starlight by dust particles (an attempt to rescue steady state) then appeared extremely contrived. The CMB really is the relic thermal radiation of the early Universe and the Universe actually was once much denser and hotter and smoother than at present.

Recall that all of the CMB photons are coming from an opaque shell at a redshift of 1000. The Solar System is moving toward one side of this shell and hence the photons detected in this direction are very slightly blueshifted. On the opposite side of the shell, the photons are similarly redshifted. In this sense the CMB establishes a preferred universal frame, and microwave radiometers can actually measure our motion with respect to this frame. This motion had already been detected in previous experiments, but COBE precisely measured the speed and direction of the Earth with respect to this universal frame – a successful variant of the old Michelson–Morley experiment attempting to measure the motion of the Earth through the hypothetical ether. It turns out that, after making the appropriate corrections for the motion of the Sun around the center of the Galaxy and the motion of the Galaxy with respect to the local group of galaxies, this local group is moving with a velocity of 620 km/s toward a large mass concentration within 100 Mpc.

Certainly the most dramatic result from COBE was the detection, announced in 1992, of the long sought-after fluctuations in the CMB temperature – the fluctuations tracing the density variations that presumably give rise to the presently observed large scale structure of the Universe. Since the discovery of the CMB, the primordial fluctuations, required if structure forms by gravitational collapse, had been the holy grail of CMB observations. Now at last, these temperature fluctuations, and, by implication, the density fluctuations, were found at about the level necessary for the formation of structure in the context of general relativity given the presence of cold dark matter. These fluctuations were seen on angular scales larger than the COBE beam size of seven degrees. This corresponds to a linear scale that would have expanded to over one thousand megaparsecs in the present Universe (comparable to the present horizon) and so would not have formed structure on any scale where it is presently observed. COBE could not resolve, by far, the fluctuations necessary for the formation of galaxies, or clusters, or even superclusters – but, by implication, these fluctuations should be present.

This major discovery led, in 2006, to the second Nobel prize awarded in connection with the CMB. John Mather of NASA, the overall leader of the COBE team, and George Smoot of the University of California at Berkeley, head of the differential radiometer group, shared this award for what has been called the beginning of precision cosmology. Certainly cosmology was becoming much more than a speculative pastime; it was evolving into an observational science which could begin to constrain fundamental theories of physics. From the point of view of the discussion here, the COBE result was totally consistent with the presence of dark matter – cold dark matter – and reinforced the emerging paradigm. But even more dramatic results were to come in the following 10 years.

9.3 What do we see in the CMB?

Recall that density fluctuations can only collapse under the influence of their self-gravity if their size exceeds the Jeans length – the distance over which a sound wave can travel in a collapse timescale. Shorter-scale fluctuations do not collapse but propagate as sound waves. In the early Universe, before a redshift of 1000 when hydrogen is totally ionized and photons and baryons are coupled as a single fluid, the sound speed in this photon–baryon fluid is effectively the speed of light, and that means that the Jeans length is comparable to the scale of the horizon, i.e., a causally connected region. All causally connected fluctuations are, in effect, sound waves – acoustic noise in the hot expanding universal baryon–photon fluid.

At the time of baryon–photon decoupling, at a redshift of 1000, when the photons are released to free-stream and be detected by us 13 billion years later, this sea of sound waves is frozen in as a pattern of CMB temperature fluctuations. The waves with the longest wavelength would be those which just enter the horizon at the epoch of decoupling – they would essentially reflect the size of the "acoustic horizon" at $z = 1000$. But we should also see higher overtones of this fundamental frequency – fluctuations that entered the horizon earlier having a wavelength only one-half, one-fourth, etc., of the horizon scale at z = 1000. Therefore, in a statistical analysis of the CMB temperature fluctuations we should see a series of peaks corresponding to this fundamental wavelength and its overtones at various angular scales.

In principle we know the absolute linear scale of the acoustic horizon at the epoch of decoupling, z = 1000 (it is dependent upon the cosmological model). So this scale is, in some sense, a standard meter stick. But from the observations of the CMB we may determine directly the angular scale corresponding to this standard meter stick. The angle subtended by a standard meter stick gives us the distance which, in this case, is also dependent upon the cosmological model. So we can take the measure of the Universe with the CMB; we can determine if the Universe is closed ($\Omega_{total} > 1$), open ($\Omega_{total} < 1$) or flat ($\Omega_{total} = 1$).

That is a lot to extract from the CMB, but even more is possible. In the context of the CDM paradigm, the fluctuations are also present in the dark matter component, but because the dark matter fluctuations are pressure less they are slowly collapsing under their self-gravity and not oscillating as sound waves. The gravitational potential wells created by these dark matter fluctuations influence the amplitude, $\delta T/T$, of the temperature fluctuations observed in the CMB. This is because, for a positive density fluctuation, the photons must climb out of the wells and suffer an additional gravitational redshift. The more dark matter, the deeper the wells. So again, in principle, the density of the dark matter component may be determined by looking at the amplitude of the various peaks, the fundamental sound wave and its harmonics, in the CMB fluctuations.

This theoretical view of fluctuations in the CMB has been developed over several decades with contributions by many scientists, beginning in 1967 with a famous paper by Ray Sachs and Arthur Wolfe, then at the University of Texas. Shortly after the discovery of the CMB and long before anyone thought about anisotropies and their consequences, Sachs and Wolfe correctly identified the primary mechanism by which large-scale density fluctuations produce corresponding anisotropies in the temperature of the CMB. The mechanism is essentially that of gravitational redshift as explained above. The subsequent theory of small fluctuations was first considered by Joseph Silk in 1967; Rashid Sunyaev and Yakov Zeldovich (1970) specifically described the imprint of acoustic oscillations on the CMB, and this was developed further in the 80s and 90s by Richard Bond, George Efstathiou, Wayne Hu, Naoshi Sugiyama, Uros Seljak and Matias Zaldarriaga. The theoretical framework was in place before observations of smaller-scale anisotropies became available, so the basic appearance of acoustic peaks in a statistical analysis of CMB anisotropies was essentially a prediction of the standard cosmological scenario with the detailed positions and shapes tracing the underlying cosmology.

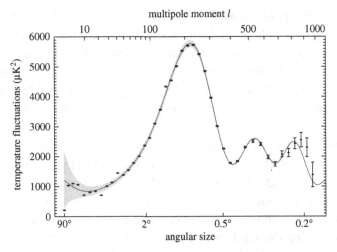

Fig. 9.3. The solid curve is the angular power spectrum of fluctuations in the CMB predicted by the concordance model of the Universe (roughly 5% baryons, 25% CDM and 70% dark energy). This is effectively the amplitude of the sound waves, in units of temperature fluctuations squared, as a function of angular scale. The points are the observations due to WMAP (NASA, WMAP science team).

This statistical analysis of expected fluctuations in the CMB is shown by the solid curve in Fig. 9.3. This is a so-called power spectrum of the fluctuations showing the amplitude of the "sound waves" on various angular scales; the solid curve is the theoretical prediction from the concordance model. The first peak corresponds to the fundamental sound waves that have just entered the horizon at the epoch of decoupling; the subsequent peaks are the higher overtones. It is immediately obvious from this plot that this series of acoustic peaks can only be seen by radiometers with resolutions of one-half degree or less. This means that COBE could not have possibly observed the acoustic oscillations but only the largest-scale anisotropies that extend out to the present horizon. Detection of the predicted acoustic peaks could only be achieved by the next generation of detectors.

9.4 Boomerang to WMAP: the age of precision cosmology

In mid-summer in the southern hemisphere the Sun never sets on the continent of Antarctica. High in the stratosphere a nearly circular pattern of east–west winds prevails so that a balloon released at one point (McMurdo station, for example) will be carried around the continent and, after some days, return to about the same point. Because Antarctica is recognized by treaty as an international zone, balloons can be released and recovered anywhere without having to worry about the complications of national boundaries and sovereign territory.

In 1999 a group of creative CMB observers took advantage of this fact and released a stratospheric balloon, Boomerang, carrying an array of sensitive detectors – in this case, bolometers which can measure tiny temperature differences – cooled to a fraction of a degree above absolute zero. This instrumentation could measure fluctuations in the CMB on angular scales of less than a quarter degree, and so could, and did, see the acoustic peaks in the CMB power spectrum.

The results of the Boomerang experiment, when published in 2000, sent shock waves through the cosmology, astrophysics and physics communities (see de Bernardis *et al.*, 2000). The predicted pattern of acoustic oscillations was clearly revealed out to the second overtone, the third peak, and the results were entirely consistent with the emerging concordance model of the Universe. From the angular size of the fundamental oscillation – the standard yard stick – the Universe appeared to be flat ($\Omega_{total} = 1$) to high precision, and, from the amplitudes of the peaks, the total matter content of the Universe was constrained to be around 30%, with only 5% in baryons. This result on baryons is entirely consistent with that deduced from observations of the light elements (helium, deuterium, etc., combined with the theory of primordial nucleosynthesis) – a striking agreement between methods based upon very different processes occurring when the Universe was at an age of a few minutes and at 300 000 years.

The CMB anisotropy pattern is not very sensitive to the contribution of dark energy, but if $\Omega_{total} = 1$ with only 30% of this in non-relativistic matter, then the rest must be dark energy. The observations – the positions and amplitudes of the acoustic peaks – agreed perfectly with that predicted by the concordance model. At about the same time, other balloon-borne (MAXIMA) and ground-based (e.g., TOCO, VSA) observations were providing results entirely consistent with Boomerang. But a phenomenal improvement in the precision of CMB observations, and the derived cosmological parameters, came with the launch of WMAP.

The "Wilkinson microwave anisotropy probe" (named in honor of the late Dave Wilkinson of Princeton, one of the originators of the project) is a satellite-based differential radiometer. It was launched in September 2001 and placed at a very special position – the L2 Lagrangian point on the opposite side of the Earth from the Sun. At this point the gravitational attraction to the Sun and Earth is canceled by the orbital centrifugal force, and the satellite, with adjustments, remains at this position, orbiting the Sun with the Earth. This detector has produced the best map of the CMB anisotropies on scales larger than about one-half degree; the first and second peaks in the power spectrum are seen with high precision (these are the points shown in Fig. 9.3). The solid curve is a cosmological model fit to these points (Spergel *et al.*, 2007), and the implied values of the basic cosmological parameters are:

$$H_0 = 72.4 \, \text{km/s/Mpc}; \tag{9.1}$$

$$t_0 = 13.69 \text{ billion years (age of Universe)}; \tag{9.2}$$

$$\Omega_{total} = 1.099 \pm 0.1. \tag{9.3}$$

We see that these results are consistent with a flat universe as predicted by the inflationary paradigm (see Section A5), although not matter-dominated because then the lifetime would be too small for the inferred Hubble constant. In addition, the composition of the Universe was derived: this is shown in Fig. 9.4, both at the present time and at the epoch of decoupling.

This degree of precision is really quite remarkable; in some sense it appears to answer all cosmological questions and one wonders what else is to be done. Importantly, in the context of the general relativistic models for the Universe there is a clear signature of dark matter – cold dark matter – in the peak amplitudes. Does this settle the issue of dark matter for once and all? Many involved in CMB

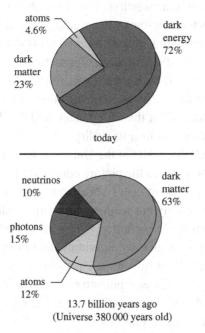

Fig. 9.4. Pie charts of the composition of the Universe at the epoch of decoupling (lower) and at the present epoch. From the amplitude of the acoustic peaks in the power spectrum the present concentration of dark matter is 23%. The dark energy, which does not dilute with the expansion of the Universe, is predominant at present, but at decoupling made no significant contribution to the total matter-energy density (NASA, WMAP science team).

observations and analysis would say so – but there is perhaps more to be said on this issue.

9.5 Reflections

The interpretation of the observed CMB anisotropies has emerged as the single most important cosmological tool. This analysis appears to tell us the age of the Universe, its present expansion rate, its geometry and its composition – all with truly remarkable precision. In a real sense, the nature of the Universe is written in the sky. It seems that there is only one remaining question: can there be any doubt?

We must bear in mind that these conclusions rest upon several assumptions and are not altogether as robust as we, at times, are led to believe. Primary among these assumptions is the validity of the standard Friedmann–Robertson–Walker cosmology and the parameters used to characterize this cosmology. Drastic changes to the Friedmann equation, resulting from new gravitational physics, have been suggested in attempts to remove the dark energy. These attempts reflect a general unease with the concordance model – a model that presents us with a universe that is quite strange in its composition. The most abundant form of matter consists of, as yet, undetected non-baryonic, weakly interacting particles originally postulated to solve the problems of the observed mass discrepancy in bound gravitational systems as well as the formation of structure via gravitational instability in an expanding universe. But even more mysterious is the dark energy – this pervasive negative-pressure fluid. The supernovae observations had already provided direct observational evidence for this component, and the detected structure of the CMB anisotropies seems to confirm its reality.

The earlier standard CDM model of the Universe was simpler; the emergence of this third component complicates the picture considerably. The dark energy is usually interpreted as the zero-point energy of the vacuum. In modern quantum field theory, the vacuum is not empty but seething with virtual particles popping into and out of existence on a timescale permitted by the uncertainty principle. The active vacuum has a non-zero energy density and should therefore gravitate – and play the role of a cosmological constant (given the equation of state of this strange fluid, it actually anti-gravitates; it produces repulsion not attraction). The problem is that, in the context of particle-physics theories this vacuum energy should be many orders of magnitude larger than that implied by the observed value of the cosmological constant: in the appropriate units of quantum gravity, Planck units, the cosmological constant has the unnaturally low value of 10^{-122}. This is seen as a fundamental problem – the most fundamental problem – by many theoretical physicists. Perhaps the vacuum does not gravitate, but this would require a profound modification of general relativity.

A more practical issue is the coincidence problem: why are we observing the Universe at a time when the density of non-relativistic matter and the density of dark energy are comparable? This is strange because the matter density dilutes with the expansion of the Universe, whereas the dark energy density does not. It is this problem that has led to the suggestion of dynamic dark energy or quintessence – a dark energy possibly associated with an additional cosmic field – a light scalar field – which also evolves with cosmic expansion possibly tracking the matter energy density.

Overall, the fact that the same rather unnatural values for the comparable densities of dark energy and matter keep emerging in different observational contexts may be calling attention to erroneous underlying assumptions rather than to the actual existence of these ethers.

Perhaps, without deeper understanding, it is too early to be overly triumphal about convergence toward a parameterized cosmology. The strange implied composition of the Universe may be signaling that the overall description of the Universe – the underlying physical assumptions – are incorrect. Nonetheless, the precision and consistency of the CMB results set a very high standard for any alternative theory.

10

An alternative to dark matter: modified Newtonian dynamics

10.1 Naive modifications of Newtonian attraction

Beginning with Zwicky, dark matter has been postulated to accommodate the disparity between the observations of large astronomical systems and the predictions of Newtonian dynamics. It has probably occurred to several readers that if Newtonian dynamics is not valid on galactic and extragalactic scales – if, for example, Newtonian gravity, or rather general relativity, breaks down in this limit – then perhaps the necessity for dark matter would vanish. As of this date (December 2012), the candidate dark matter particles have not been detected independently of their presumed gravitational effects. Therefore, the existence of dark matter remains hypothetical and is dependent upon the assumed law of gravity or inertia on astronomical scales. So it is not at all outrageous to consider the possibility that our understanding of gravity is incomplete. If a physical law, when extended to a regime where it has never before been tested, implies the existence of a medium (an ether) that cannot be detected by any other means, then it would not seem unreasonable to question that law.

It is a simple matter to cook up a recipe that explains a single aspect of the observed mass discrepancy such as flat rotation curves of spiral galaxies. For example, we may suppose that beyond some critical length scale, r_0, the attraction of gravity falls as $1/r$ rather than $1/r^2$. Specifically we could write, for the acceleration due to gravity

$$F_g = \frac{GM}{rr_0} \tag{10.1}$$

in the limit where r is much greater than r_0, a new constant of nature that must be of the order of galactic dimensions ($r_0 \approx 10\,\mathrm{kpc}$). Then if we equate this to the centripetal acceleration (due to circular motion)

$$\frac{V^2}{r} = F_g \tag{10.2}$$

we would find

$$V^2 = GM/r_0 \qquad (10.3)$$

which is constant for any given galaxy with mass M. Thus, flat rotation curves beyond r_0 are explained.

But there is more galaxy phenomenology that must be accommodated. For example, the Tully–Fisher law is the well-known power-law correlation between the luminosities of spiral galaxies and their rotational velocities $L \propto V^\alpha$. The modified gravity law suggested here would imply that $\alpha = 2$ assuming that the mass-to-light ratio in spiral galaxies does not vary systematically with luminosity or rotation velocity (it apparently does not when the luminosity is measured in the near-infrared emission typical of the dominant stars). But observations imply that α is larger than three. So such a simple modification misses a conspicuous aspect of galaxy phenomenology.

And there is more. Any modification attached to a length scale, a critical distance beyond which Newtonian attraction is modified, would imply that the mass discrepancy (the discrepancy between the luminous and dynamical masses) should be larger in larger galaxies. But this is definitely not the case. As we have seen in the previous discussion, there are very small galaxies (for example, the dwarf spheroidals in the neighborhood of the Milky Way) with large discrepancies (much dark matter) and large bright spiral galaxies (such as UGC 2885 observed by Rubin *et al.* in 1980) with small discrepancies (little dark matter). If anything, the magnitude of the discrepancy seems correlated with surface brightness and not size.

Any ad hoc modification which explains one aspect of galaxy phenomenology but gets others wrong is hardly well-motivated.

10.2 MOND

From 1980 to 1982, Mordehai (Moti) Milgrom, a young Israeli physicist at the Weizmann Institute, was on sabbatical leave at the Institute for Advanced Study in Princeton. Milgrom had worked on astrophysical problems for several years – galactic X-ray sources, models for compact radio sources, the definitive model for SS433, the bizarre precessing jet-like object in the Galaxy – and his work was characterized by creative insight and originality combined with a careful attention to the experimental facts. At Princeton he was contemplating non-dark matter alternatives to the mass discrepancy in astronomical systems; he was impressed not only by the fact that rotation curves of spiral galaxies appeared to be asymptotically flat, but also by the existence of the power-law correlation between galaxy luminosity (and presumably luminous mass) and the rotation velocity ($L \propto V^\alpha$); the exponent α was uncertain at the time but was within the range 2.5 and 5.0, clearly inconsistent with $\alpha = 2$ as required by a simple modification of Newtonian attraction such as that given by eq. 10.1.

Milgrom realized that a modification attached to an acceleration scale could accommodate both of these observational facts. He realized further that this could be viewed as a modification of Newtonian gravity or inertia; that is to say, either the law of gravitational attraction or the response of particles to the imposed force differs from that postulated by Newton. Milgrom wrote down a very simple modification of Newtonian dynamics that has become known as MOND ("modified Newtonian dynamics"). As an alternative to dark matter, MOND has proved to be extremely resilient over the past 25 years because it explains many of the systematics and details of the mass discrepancy, particularly in galaxies, and predicts trends that have subsequently been discovered.

Let us view Milgrom's hypothesis as a modification of Newton's second law, the law of inertia. When we apply a constant force F to an object with mass m, as we have all learned in high-school physics, the object exhibits a constant acceleration a. This is neatly summarized by the famous formula

$$F = ma. \tag{10.4}$$

The mass m appearing in this formula is known as the inertial mass.

Milgrom proposed that, while the original second law is entirely adequate when describing objects with accelerations comparable to those encountered on Earth or in the Solar System, in the limit of very low accelerations, such as that in the outer Galaxy – accelerations lower than some fundamental acceleration a_0 of the order of 10^{-10} m/s^2 – this law should be modified to read

$$F = ma^2/a_0. \tag{10.5}$$

So the acceleration is no longer proportional to the applied force but to the square-root of the force. The complete expression relating force and acceleration would read

$$F = ma\,\mu(a/a_0) \tag{10.6}$$

where μ is a function that interpolates between the two regimes: $\mu(a/a_0) = 1$ in the limit of large accelerations ($a \gg a_0$) recovering the familiar Newton's law, and $\mu(a/a_0) = a/a_0$ in the limit of small accelerations ($a \ll a_0$) which yields the modified dynamics (eq. 10.5).

Newton's second law and the modified second law (for a particular choice of μ) are illustrated in Fig. 10.1; this is a plot of the acceleration resulting from an applied force (per unit mass) in both schemes. The claim is that galactic and extragalactic dynamical phenomena are explained by the divergence of the true relation, MOND, and the Newtonian relation at low accelerations ($a \ll a_0$). Applying this formula

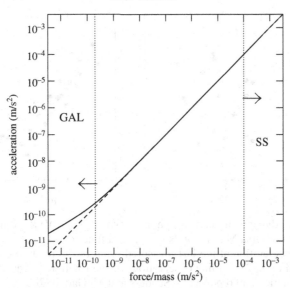

Fig. 10.1. MOND as a modification of Newton's second law. On this logarithmic plot the acceleration resulting from an applied force (per unit mass) is shown for Newtonian dynamics (dashed curve) and for modified Newtonian dynamics (solid curve). These only differ significantly below an acceleration of 10^{-10} m/s^2. The indicated region on the right corresponds to accelerations in the Solar System and that on the left to accelerations in the regions of bright galaxies. Solar-System accelerations are deep in the Newtonian regime, but galaxy-scale accelerations are typically in the modified regime. Here it is assumed that $\mu(x) = x/(1+x)$.

(eq. 10.6) to circular motion ($a = V^2/r$) in a gravitational field of a point mass M where the acceleration has fallen below a_0, we find

$$\frac{1}{a_0}\left[\frac{V^2}{r}\right]^2 = \frac{GM}{r^2} \qquad (10.7)$$

which reduces to

$$V^4 = GMa_0. \qquad (10.8)$$

In other words, the rotation curve of a galaxy (or any object) is asymptotically flat far from the mass distribution and, in general, the mass of galaxies is proportional to the fourth power of this constant asymptotic velocity.

The basis of MOND is that discrepancies in galaxies (or in any astronomical system) should appear at low accelerations, not at large distances. This is immediately testable with a homogenous sample of spiral galaxies all at about the same distance – such as the Ursa Major sample observed in neutral hydrogen at Westerbork WSRT by Marc Verheijen. Fig. 10.2 is a plot of the Newtonian dynamical mass-to-light ratio for the Ursa Major spirals plotted on the left as a function of galaxy size, defined by the last measured point of the rotation curve, and, in the right

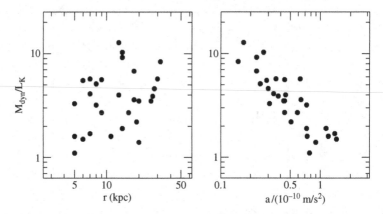

Fig. 10.2. The dynamical mass-to-light ratio (where $M = V^2R/G$ and L is in the near-infrared) of Ursa Major spirals shown as a function of galaxy size (left) and centripetal acceleration at the last measured point of the rotation curve (V^2/R). This is based upon 21-cm line observations of the rotation curves at Westerbork by Marc Verheijen and reported by Sanders and McGaugh (2002).

panel, as a function of centripetal acceleration at the last measured point (V^2/R). The Newtonian mass is estimated using the simple formula $M = V^2R/G$. Normal spiral galaxies have a M/L of about one in the near-infrared, and any value much larger than this would suggest a discrepancy. We see immediately that there is no correlation of M/L with the size of the galaxies; some small galaxies have large discrepancies and vice versa. But when M/L is plotted as a function of centripetal acceleration, there does appear to be a trend. The smaller the acceleration, the larger the discrepancy and this discrepancy becomes obvious at accelerations below 10^{-10} m/s^2. It does indeed seem that the Newtonian dynamical mass-to-light ratio is inversely correlated with acceleration as Milgrom proposed (note that Milgrom's proposal preceded this data by 15 years).

If MOND is expressing physical law, then flat rotation curves and the $M \propto V^4$ relation must be absolute – there can be no exceptions. Moreover, the true Tully–Fisher law is a relation between the baryonic mass of a galaxy and the constant asymptotic rotation velocity – the rotation velocity at a large distance from the galaxy. Looking back at Fig. 8.3 we see the Tully–Fisher law for the Ursa Major cluster sample where the luminosity is measured in the near-infrared (most nearly proportional to the stellar mass), and the rotation velocity is that beyond the visible disk. The straight line on this logarithmic plot is the expectation for MOND. Assuming a reasonable mass-to-light ratio for the stars in these galaxies (M/L $\approx$ 1) and shifting the line so that it fits the observations, sets the new fundamental physical constant of the theory, $a_0 = 10^{-10}$ m/s^2. Provocatively, this is comparable (within a factor of six or seven) to cH_0, the speed of light multiplied by the Hubble parameter. In other words, if a particle accelerates at a_0 then it will reach a velocity about equal to the speed of light in the lifetime of the Universe (the Hubble time). This coincidence with cosmology is highly suggestive: perhaps MOND represents the effect of cosmology on local particle dynamics in the limit of low acceleration. Perhaps a_0 varies with cosmic time as does the Hubble parameter, or perhaps it does not vary, like a cosmological constant.

When Milgrom's first papers appeared in 1983, the actual value of the exponent in the Tully–Fisher relation was not at all certain. But after another decade of rotation-curve observations, in which the asymptotic constant rotation velocity and the near-infrared luminosity of many galaxies were measured, it became clear that the exponent is actually *four*, as required by such an acceleration-based modification. It has also become evident (thanks largely to the work of Stacy McGaugh), that the true relation is between the rotation velocity and the baryonic matter content of a galaxy – a matter content that includes the contribution of gas as well as stars (McGaugh *et al.* 2000).

MOND goes further and predicts general trends in the nature of the discrepancy in astronomical systems. The acceleration parameter may be written as a surface density:

$$\Sigma_0 = a_0/G$$

(both acceleration and surface density are proportional to M/r^2). This surface density is about 700 $M_\odot/\mathrm{pc}^2$ or 0.15 g/cm^2, comparable to a dozen pages of this book. Whenever an astronomical object has a surface density less than this critical value, it is in the low-acceleration regime – the MOND regime; a higher surface density means Newtonian dynamics. This leads to a very definite and inescapable prediction: in so far as surface brightness reflects surface density, low-surface-brightness objects should exhibit a large mass discrepancy, when considered with Newtonian dynamics, and high-surface-brightness objects should exhibit a small mass discrepancy. This would seem to explain why diffuse faint objects such as dwarf spheroidal galaxies or low-surface-brightness spiral galaxies require, with Newtonian dynamics, considerable dark mass within the visible object. It would also account for the fact that compact bright objects such as globular star clusters, bright elliptical galaxies, or high-surface-brightness spirals (such as those observed by Rubin and collaborators) seem to require very little dark matter within the optical image. Viewed in terms of MOND we can understand why the maximum-disk works so well for bright spiral galaxies (see Chapter 5). This result going back to Schwarzschild, Kalnajs and Kent now becomes comprehensible.

At the time that MOND was proposed, the claim that low-surface-brightness galaxies should have a large discrepancy was, in fact, a prediction. Very few spiral galaxies with low surface brightness (LSB galaxies) had been discovered at that point (they are generally fainter than the night sky and thus hard to find). But since that time, a large population of LSB galaxies has been discovered and observed in the 21-cm line and optical emission lines. Without exception, a large discrepancy is present within the optical disk of these objects; in traditional language, these galaxies require a significant mass of dark matter within the visible image. This is not an evident prediction of the CDM paradigm – a point that has been often emphasized by McGaugh.

With a mass-to-light ratio of three (in blue light) Σ_0 would correspond to a surface brightness of about 200 $L_\odot/\mathrm{pc}^2$. This fiducial surface brightness is comparable to Freeman's characteristic value (Chapter 4); it is the upper limit on the observed central surface brightness of disk galaxies. This is quite interesting because when the surface brightness is higher than this value, the disk would be essentially Newtonian, and we have seen (Chapter 3) that rotationally supported Newtonian disks tend to be unstable (recall that this was the original motivation of Ostriker and Peebles for proposing massive spheroidal dark halos). In the present context this could well explain why we see no spiral disk with a higher surface brightness; they cannot exist because they are not stable. Freeman's law finds explanation in MOND.

The existence of a critical surface density has an additional consequence for the general form of galaxy rotation curves. A high-surface-brightness galaxy is Newtonian within the optical image of the galaxy, so the expectation is that a rotation curve beyond the visible image should decline to its final asymptotic value given by eq. 10.8. On the other hand low-surface-brightness galaxies are deep within the MOND regime, so we would expect that the rotation curves of such objects should slowly rise to the asymptotic value. In other words, there should be a systematic difference between the rotation curves of HSB and LSB galaxies. This was one of Milgrom's original predictions, and, just such a difference was identified by Casertano and van Gorkom some years afterwards (1991). This difference is well illustrated by Fig. 8.2, which shows the rotation curves of LSB and HSB galaxies along with the MOND-predicted rotation curves.

10.3 MOND and hot galaxies

MOND, as physical law, should also be relevant to galaxies and other systems that are supported, not by rotation, but by random motion (i.e., pressure), such as elliptical galaxies or clusters of galaxies or globular star clusters. To consider the structure of such objects we must solve the equation of hydrostatic equilibrium using Milgrom's and not Newton's formula for the acceleration (first done by Milgrom in 1984). To solve the equation, we need an additional assumption about how the random velocity is related to the density – an assumption such as Emden's

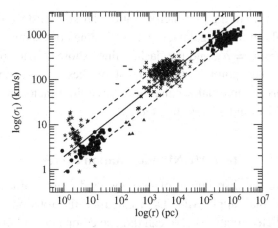

Fig. 10.3. The line-of-sight velocity dispersion plotted against the characteristic radius for pressure-supported astronomical objects. The star-shaped points are globular clusters, the solid round points are massive molecular clouds in the galaxy, the triangles are dwarf spheroidal galaxies, the crosses are elliptical galaxies and the squares are clusters of galaxies. The solid line shows the relation $\sigma^2/r = a_0$. The references for the observations are given by Sanders and McGaugh (2002).

power-law relation (see Chapter 7). If we make the isothermal assumption (the random velocity of the stars is constant) then we find that, unlike the Newtonian isothermal sphere, the MOND isothermal sphere has a finite mass. Moreover, this mass is proportional to the fourth power of the velocity dispersion, $M \propto \sigma^4$, where σ is usually measured by the width of the stellar spectral lines. There is such a relation observed for elliptical galaxies; it was discovered in 1976 by Sandra Faber and Robert Jackson (UC Santa Cruz) – the Faber–Jackson relation. MOND would imply that this relation should apply to all near-isothermal pressure-supported systems. Hence MOND would explain why an object with a velocity dispersion of 5 km/s has the mass of a globular cluster (10^5 $M_\odot$), an object of 100 km/s the mass of a galaxy (10^{11} $M_\odot$) or an object of 1000 km/s the mass of a cluster of galaxies (10^{14} $M_\odot$).

The MOND isothermal sphere resembles the Newtonian isothermal sphere out to a radius where the gravitational acceleration falls to the critical value, a_0; this would be $r_m = \sqrt{GM/a_0}$. At larger distances the density of the MOND sphere declines more rapidly (as $1/r^4$) because of the larger effective gravitational force; thus r_m becomes the effective radius of the system (R). This means that the typical acceleration of a particle inside a MOND isothermal sphere, estimated by σ^2/R, should be of the order of a_0. What do the observations reveal about this?

Fig. 10.3 shows the line-of-sight component of the velocity dispersion plotted against the typical radius of pressure-supported objects ranging from sub-galactic objects to the giant clusters of galaxies. The meaning of the points is indicated in the figure caption, but the plot includes globular star clusters, dwarf spheroidal galaxies and bright elliptical galaxies. The solid line corresponds to fixed internal acceleration of $\sigma^2/R = a_0$, and the dashed lines show a factor of three variation about this value. The point is that in most of these near-isothermal pressure-supported systems the internal acceleration is within a factor of three of a_0. It is unclear how CDM would address this fact.

10.4 MOND and rotation curves

If we can observe the distribution of baryons, and there is no matter other than baryons, then we may apply MOND to calculate the rotation curve of a spiral galaxy. This predicted rotation curve can then be compared with the observed rotation curve. In this sense, rotation curves should constitute a strong test of MOND. The dark matter hypothesis is rather immune to tests of this sort because when there is an unseen component of the mass distribution, essentially any observed rotation curve may be matched by adjusting the distribution of dark matter (a bit like explaining planetary motion with unseen crystal spheres). There are some constraints imposed by the universal density profile for CDM halos and the relationship of halo parameters, such as concentration, via cosmology, to halo mass, but these are fairly weak.

The procedure is really very similar to that followed by Schwarzschild (Chapter 2) or Kalnajs (Chapter 4) in predicting the Newtonian rotation curves of spiral galaxies. We assume that the light traces the mass (constant mass-to-light ratio) and that this mass is in a thin disk. We then add in the directly observed contribution of the neutral hydrogen (plus the primordial helium). Given this mass distribution we use Milgrom's formula to calculate the effective force and then the rotation curve. We compare the calculated curve with that observed and adjust the mass-to-light ratio of the visible disk to achieve the best match. The disk M/L is therefore the only free parameter in this process; we are not allowed to readjust the acceleration parameter, a_0, because it is a fundamental constant.

This procedure has been carried out for about 100 galaxies, and Fig. 10.4 shows a familiar example, NGC 2403. The Newtonian rotation curve of the detectable baryonic components, stars and gas, is shown by the dotted line; the solid curve is that computed from the Newtonian force modified by the MOND algorithm. The match between the observed rotation curve, and that predicted by MOND is evident (here $a_0 = 10^{-10}$ m/s^2 as usual). It is also evident that the rotation curve beyond the disk is flat and featureless; in particular, the disk–halo conspiracy that arises in the context of dark-halo models (never a distinct feature in the rotation curve for a halo) becomes quite irrelevant in the context of MOND. The required mass-to-light ratio (0.9 in solar units) is entirely consistent with that expected from the population of

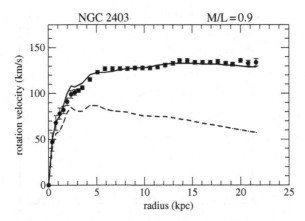

Fig. 10.4. The rotation curve of NGC 2403. The points show the observed rotation curve. The dashed curve is the Newtonian rotation curve of the detectable baryonic matter – stars and gas – with an assumed mass-to-light ratio for the stellar disk of 0.9 in solar units. The solid curve is the total rotation curve determined with Milgrom's formula.

stars in such a spiral galaxy. This is true in general; for the Ursa Major sample, the M/L values required by MOND match the overall trends predicted by models of the mix of stars in spiral galaxies: blue galaxies (low B-V) with ongoing star formation are found to have systematically smaller M/L values than red galaxies (high B-V) with low star-formation rates (MOND has no way of "knowing" that this should be the case).

It even appears that details in the rotation curves are matched by the predicted MOND rotation curves. A more striking example of this is provided by the rotation curve of the low-surface-brightness galaxy, UGC 7524, shown in Fig. 10.5. The top panel is the surface density in this galaxy for the stars (assuming constant M/L) and gas corrected for primordial helium. The bottom panel shows the corresponding Newtonian rotation curve (dashed curve); the points are the observations and the solid curve is the MOND rotation curve determined from the Newtonian curve after processing with Milgrom's formula. Here we see that features in the observed baryonic mass distribution have their counterparts not only in the predicted Newtonian rotation curves, as expected, but also *in observed rotation curves*.

As often discussed by the radio astronomer Renzo Sancisi from a strictly observational point of view, this detailed matching of features in the Newtonian and observed rotation curves is generally true in spiral galaxies, even in the presence of a large discrepancy between the detectable and Newtonian dynamical mass. This would seem rather strange in the context of dark matter; it would seem as though the distribution of dark matter is extremely responsive to the distribution of visible matter – a bit like the tail wagging the dog. The minimal implication is that

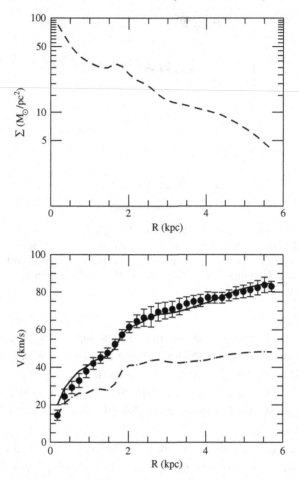

Fig. 10.5. The top panel shows the surface density of the detectable baryonic matter, stars and gas, plotted as a function of radius in the disk of the low-surface-brightness galaxy UGC 7524. The lower panel is the observed rotation curve (points with error bars), the Newtonian rotation curve of the detectable disk (resulting from the surface density distribution above) and the total MOND rotation curve (solid curve). Note that the rather abrupt increase in rotation velocity near 2 kpc corresponds to enhancements in the observable matter density at these radii. Based upon observations at Westerbork by Swaters (1999).

the visible and dark components are very closely coupled. But dark matter and baryonic matter are two very different sorts of fluids. Baryonic matter is susceptible to hydrodynamical effects: it may be shocked and removed in collisions; it may be blown away by supernovae. But the only force influencing dark matter is, presumably, gravity; it is dissipationless.

Such considerations are problematic for dark matter, at least on the scale of galaxies. With MOND, the precise matching of structure in the rotation curve with structure in the baryonic matter distribution is the expected result. This demonstrates that MOND may be viewed as an algorithm that allows one to predict the distribution of force in an astronomical object from the observed distribution of baryons. The fact that such an algorithm exists, and that it works, constitutes a considerable challenge for standard CDM.

10.5 The problem of clusters

MOND appears to explain the magnitude and nature of the discrepancy on galactic and sub-galactic scales. Applied to the dwarf spheroidal galaxies in the near neighborhood of the Milky Way – systems that are heavily dominated by dark matter in the context of Newtonian dynamics – the prescription returns mass-to-light ratios that are generally consistent with those expected for a normal stellar population. In high-surface-brightness systems (systems with high internal accelerations) such as the bright elliptical galaxies, various dynamical tracers of the mass distribution (e.g., kinematics of planetary nebula) indicate very little dark matter within the optical image of the galaxy – completely consistent with MOND expectations. But what about larger scales? The discrepancy was first identified in the rich clusters, so one might hope that a non-Newtonian alternative would also explain the unexpectedly high random velocities of galaxies (or high gas temperatures) in clusters without dark matter.

A powerful tracer of the total mass distribution in clusters is provided by the observed density and temperature distributions of the hot gas (Chapter 7). By applying the Newtonian equation of hydrostatic equilibrium we determine the distribution of total mass – baryonic plus dark. The same can be done with MOND, by substituting Milgrom's formula instead of Newton's for the gravitational force.

Fig. 10.6 shows what is found by the two theories. On this logarithmic plot the dynamical mass is plotted against the observable mass in gas and visible galaxies; the left panel is with Newtonian dynamics, the right is with MOND. The straight line would correspond to no discrepancy. Here we see that in the Newtonian analysis the total dynamical mass is six or seven times larger than the observable baryonic mass. With MOND, however, the discrepancy is reduced to a factor of two or three on average. So MOND reduces the discrepancy but it does not remove it.

Is this a falsification of MOND? Actually, no. If MOND predicted less mass than is actually seen in clusters, it would be a definitive falsification. We may always find more mass but we cannot make directly observed mass in the form of hot gas or stars vanish. But this result is certainly a challenge for MOND because this proposed alternative to dark matter still requires undetected (as yet) matter – dark matter – in clusters of galaxies. The optical, X-ray and lensing observations of

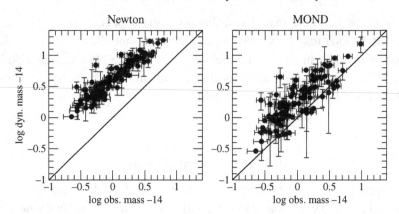

Fig. 10.6. The panel on the left is a log–log plot of the Newtonian dynamical mass against the directly observed baryonic mass (mostly in the form of hot gas). The units are 10^{14} $M_\odot$. The panel on the right is the same for the MOND dynamical mass. Note that MOND reduces the discrepancy but does not remove it.

the famous "Bullet cluster" (Chapter 7 and rear book cover) – the two colliding clusters of galaxies in which the dominant baryonic component, the gas, is clearly separated from the major mass concentrations as traced by gravitational lensing – are presented as compelling evidence for dark matter and against MOND. It is a problem for MOND, but not an additional problem. The significance of the Bullet is that unobserved matter, whatever it is, behaves like the stars and not like the hot diffuse gas – it is dissipationless.

There is more than enough baryonic matter in the universal mass budget to provide the unobserved component, but it should be in some dissipationless form – "massive compact halo objects" (MACHOs) or small dense clouds, for example. It could also be that the undetected matter in clusters consists of neutrinos. Non-baryonic matter certainly exists in the Universe in the form of ordinary neutrinos; these elusive particles have mass (greater than 0.05 eV) and they are as numerous as the CMB photons. If the mass were as large as 1.5 eV, they could provide the missing mass in clusters but they would not accumulate in galaxies (the packing velocity would be too high). The possibility of neutrinos as cluster dark matter will be ruled out (or not) by experiments presently underway to measure the neutrino mass, but, for now, the remaining dark matter in clusters constitutes the greatest observational challenge for MOND.

10.6 Relativistic MOND: TeVeS

MOND embodies and unifies a range of phenomena, primarily on the scale of galaxies – phenomena, such as the preferred surface density in spiral galaxies, flat rotation curves, the Tully–Fisher law and the Faber–Jackson relation – all of which would apparently be unrelated in the context of CDM. However, MOND remains silent on issues of gravitational lensing, cosmology, structure formation, the CMB. And these are no longer purely speculative problems; gravitational lensing is now an effective astronomical tool in mapping the mass distribution in astronomical systems, and cosmology has become a solid observational science in the past 25 years. This silence is due to the fact that the theory in its original form lacks a relativistic extension; it is clearly incomplete.

This deficiency was certainly realized by Milgrom and other supporters of MOND. One of those early supporters was the eminent physicist Jacob Bekenstein. Bekenstein, who is an expert on the theory of black holes, has also given considerable thought to alternative theories of gravity – in particular, scalar–tensor theories. A word of explanation is necessary here. General relativity is a field theory of gravity. The field that "carries" the gravitational force is a mathematical object called a tensor with 10 independent parts. So in a sense, there is not just one field describing gravity, as in Newtonian theory – there are 10 (in general relativity this tensor happens to be the so-called metric tensor of space–time and this gives the theory its powerful geometric interpretation). Since the time of Einstein, people have thought about alternatives, or enlargements, to general relativity; one obvious enlargement was to add an additional field – a scalar field which has only one component. Scalar–tensor theories were developed by the German physicist Pascual Jordan in the 1940s and, 20 years later, carried further by Carl Brans and Robert Dicke at Princeton.

In 1983, Bekenstein, then at Beersheva in Israel, became interested in Milgrom's proposal and thought that it might be possible to realize MOND as a scalar–tensor theory of gravity. Bekenstein and Milgrom collaborated in work published in 1984 in which they considered MOND as a modification of gravity, not inertia. They wrote down a non-relativistic field theory of MOND that successfully addressed several theoretical problems in Milgrom's original formulation (such as non-conservation of momentum). And further, in an appendix, they demonstrated how such a theory could be made relativistic and so address problems of light-bending and cosmology, at least in principle. The idea, termed AQUAL or "aquadratic Lagrangian" theory, was a scalar–tensor theory but one in which the strength of the scalar coupling to ordinary matter depended upon the strength of the scalar field force. It was an acceleration-based modification; at low accelerations the scalar field affected the motion of a particle more strongly. It may also be considered as a theory involving an additional force that gives rise to MOND phenomenology, "a fifth force" as is illustrated in Fig. 10.7.

There were some theoretical problems with this original suggestion – such as faster-than-light propagation of waves – which kept Bekenstein and others

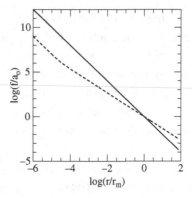

Fig. 10.7. This is a log–log plot of the usual Newtonian $1/r^2$ force (solid curve) and an anomalous "MOND" force (dashed curve) as a function of distance from a point mass M for AQUAL scalar–tensor theories. The force is given in units of the MOND acceleration parameter a_0 and the distance is given in terms of that critical distance $r_M = \sqrt{GM/a_0}$ where the acceleration has fallen to a_0. We see that the $1/r$ MOND force exceeds the usual Newtonian force not beyond a critical distance but below a critical acceleration. At smaller distances, the MOND force is also $1/r^2$ – necessary in order to match the phenomenology of planetary motion in the inner Solar System.

(including Milgrom and myself) busy for another 20 years. The biggest problem, and one with a relevance to actual observations, turned out to be that presented by gravitational lensing. A scalar field should couple to a massive particle in such a way that its motion is independent of its composition (the coupling of a field to a particle describes how much a field of a given strength will affect the motion of that particle). That means that, in a vacuum, a ping-pong ball should fall at the same rate as a bowling ball, as shown by Galileo four centuries ago (although not with ping-pong and bowling balls). Einstein elevated this "universality of free-fall" to the level of a principle, the principle of equivalence, which became the basis of general relativity. However, the traditional way of including this condition in scalar–tensor theory, the so-called conformal coupling used by Jordan and by Brans and Dicke, implies that the scalar field does not interact at all with relativistic particles such as photons. In other words, there should be no additional deflection of a photon due to the presence of the scalar field. This means that gravitational deflection of photons by a giant cluster of galaxies should be only that expected from the detectable baryonic matter. The prediction would be the mass determined by gravitational lensing should be much less than the mass determined by the Newtonian virial theorem. This is a striking contradiction to the observations. By 1990 it was clear that the lensing mass of clusters was comparable to the virial mass.

Bekenstein and I had realized that a more complicated coupling of the scalar field to matter was required in order to combine the universality of free-fall with the appropriate deflection of photons – a "disformal" coupling (Bekenstein and Sanders, 1994). In 1997, while contemplating a historical class of modified gravity theories, "stratified theories", I decided that this disformal coupling could best be achieved by adding an additional field – a different sort of field called a vector field that has four components in four-dimensional space–time. I wrote down a trial theory that included these three sorts of fields: tensor, vector, scalar. The difficulty with my idea was that the vector was put in by hand; it was a non-dynamical field, which means that it can affect the motion of particles but is not itself affected by the distribution of particles. This is totally at odds with the spirit of relativity.

In 2004, Bekenstein, in a masterful piece of work, repaired this deficiency and introduced a dynamical vector field – one that is determined via its own field equation from the actual mass distribution. What emerged was TeVeS – a fully dynamical "tensor–vector–scalar" relativistic theory of MOND. Finally there was a consistent relativistic theory underlying MOND. For years the absence of such a theory was used as an argument against MOND, but Bekenstein demonstrated that it can be done. Therein lies the principal importance of the theory.

Unlike general relativity, TeVeS is an entirely inductive theory, i.e., it was constructed from the bottom-up with the various elements added in response to perceived anomalies or in order to match phenomenology. It is designed to produce the general relativistic relation between the total weak field force (including that resulting from the scalar field) and the deflection of photons. With respect to general relativity it contains two additional fields, three additional parameters (one of which is the acceleration parameter a_0, put in by hand), and one free function, i.e., a function whose form is not specified by any a priori considerations but is set in order to reproduce MOND phenomenology. The theory has been criticized on the grounds of being overly complicated and unaesthetic. Probably it should be seen as an "effective theory" – one that is not fundamental itself but emerges from a more basic theory. There are still too many adjustable elements (such as the form of the free function in the cosmological limit) to specify a "standard" TeVeS cosmology, but initial work by Constantinos Skordis and others (2006) does indicate that the theory can confront problems of structure formation and CMB anisotropies.

10.7 Summing up: MOND vs. dark matter

MOND in its most basic form may be seen as an algorithm which predicts the distribution of force from the observed distribution of baryonic matter. Regardless of its theoretical basis, the fact that this algorithm is successful on the scale of galaxies may be viewed as a falsification of CDM, because this is not something that standard dark matter can naturally do. Problems such as the density cusp in

CDM halos (the NFW halos) or the large number of predicted satellite halos, are secondary.

It may be argued that MOND describes a systematic but poorly understood relationship between the visible and dark components in galaxies, but this would be seem to require a dark–visible coupling which is totally at odds with the proposed nature of CDM. Baryons behave quite differently than dark matter: they dissipate and collapse to the center of a system; they are left behind in collisions between galaxies or between clusters; they are blown out by supernovae. It is difficult to comprehend the intimate connection between baryons and dark matter particles implied by the phenomenology of rotation curves.

Moreover, there is the ubiquitous appearance of $a_0 \approx cH_0$ in the phenomenology. CDM halos embody a density scale but not an acceleration scale. How then does CDM account for the fact that a_0 is the acceleration at which the discrepancy appears in galaxies, that a_0 sets the normalization of the Tully–Fisher relation for spiral galaxies and the Faber–Jackson relation for elliptical galaxies, that a_0 is the characteristic internal acceleration of spheroidal systems ranging from molecular clouds to clusters of galaxies and that a_0 defines a critical surface density which appears as the upper limit for spiral disks (Freeman's law)? All of these phenomena would seem to require a separate explanation in terms of CDM; with MOND they are unified.

What would constitute a verification of MOND (in so far as theories can be verified)? Theories, like TeVeS, do, inevitably, predict new phenomena on scales that are very different than of the galactic and extragalactic arena for which MOND was conceived. For example, the vector field in TeVeS establishes the cosmological frame as a preferred frame: the internal motion of a gravitating system, like the Earth–Moon system, is slightly different in a frame moving with respect to the cosmological frame, as is the Solar System. It may be possible to observe small local "ether-drift" effects such as a slight stretching of the orbit of the moon in the direction of the drift. Moreover, almost any multi-field theory, such as TeVeS, predicts a very slight non-Newtonian behavior in the outer Solar System where the accelerations are low.

This is quite interesting in view of a curious anomaly found by the Pioneer spacecrafts. Pioneer 10 and Pioneer 11 were launched in 1972 and 1973 in order to explore the outer Solar System and are on trajectories that will carry them out of the Solar System into interstellar space. Both spacecrafts weigh about 260 kg and are powered by onboard nuclear reactors. When these spacecrafts had moved beyond a solar distance of about 2.8 billion kilometers (beyond the orbit of Uranus), the scientists at the Jet Propulsion Laboratories in California (a group led by John Anderson) noticed something strange. Both objects did not move exactly as they should in the pure inverse-square gravity force due to the Sun. They appeared to be moving more slowly; in fact, they appeared to be decelerating due to an additional constant force directed in toward the Sun. The magnitude of this anomalous acceleration is about 8.7×10^{-10} m/s^2, almost exactly equal to cH_0 and about six or seven times larger than Milgrom's a_0. Anderson's group (1998) considered every possible prosaic explanation they could think of for this effect: anisotropic radiation of heat generated by the onboard reactors, gas leaks of the onboard propellent, anisotropic re-radiation of absorbed solar energy. Nothing seemed to work.

So at present the Pioneer effect is without conventional explanation. This is of considerable interest because it is the very first indication that gravity or dynamics is not strictly Newtonian in the outer Solar System, that our understanding of gravity or Newtonian dynamics may be incomplete on the scale of the Solar System, and that the deviation appears at low accelerations. The magnitude of the anomalous acceleration, being so close to the MOND acceleration, is tantalizing.

In general, such effects, on a scale vastly smaller than that of galaxies, may be considered as the "holy grail" of modified gravity alternatives to dark matter. Dark matter itself has a different holy grail – direct detection.

11

Seeing dark matter: the theory and practice of detection

11.1 Non-gravitational detection of dark matter

Anyone who thinks objectively about the concordance model of the Universe must surely be concerned that 80% of the matter content of the Universe has never been detected independently of its global gravitational effects in large, and generally distant, astronomical systems. It is a bit as though Neptune had never been discovered after having been postulated to account for the anomalous motion of Uranus. Not only has the non-baryonic particle matter never been seen (it is of course, dark), but we have no definite idea of the identity of these putative cold dark matter particles. As discussed in Chapter 6, there are no known "standard-model" particles that fit the bill – they must be electrically neutral, stable and slow moving (cold). But reasonable extensions of the standard model of particle physics do provide a number of candidates, and, of these, perhaps the most well-motivated is the LSP, the "lightest superpartner" that should exist in the context of the theory of supersymmetry.

To summarize the discussion in Section 6.4, the basis of supersymmetry is that every known particle has a partner that differs by a half-integral spin; for example, the partner of the spin 1 photon is the spin 1/2 photino. This proposed symmetry between integral spin particles, bosons, and half-integral spin particles, fermions, rather successfully confronts a number of theoretical problems in physics, although, so far, there is no direct experimental verification of supersymmetry. But the theory does, in effect, double the number of possible particles. Most of these hypothetical particles are heavy and unstable; they decay into more familiar standard-model particles and lower-mass superpartners soon after the instant of the Big Bang. There is, however, the lowest-mass superpartner (LSP) that should be stable and that will be present with some abundance as a relic of the Big Bang. The primary supersymmetric candidate is the neutralino, which is a mixture of neutral supersymmetric (half-integral spin) partners of the photon, the Z boson and the

150

Higgs boson: the photino, the zino and the Higgsino. The generic term for such a weakly interacting massive particle is WIMP.

The estimated abundance of this primordial relic provides the strongest argument in favor of the LSP as the dark matter particle. As the Universe expands and cools, the average energy of particles and photons drops below the rest-mass of the LSP, perhaps 100 GeV, at about 10^{-10} seconds after the instant of creation. Then the remaining LSPs and their anti-particles begin to annihilate one another until the annihilation rate decreases below the expansion rate of the Universe (the neutralino is actually its own anti-particle). This leaves a relic of such particles that persists until the present epoch and supposedly provides the dark matter required for structure formation as well as the unseen mass component of astronomical systems. As we saw, the relic abundance depends upon the annihilation cross section and not directly upon the mass of the particles. If that cross section is of the order of that for weak interactions – a reasonable supposition – then that relic abundance, Ω_X, will be of the order of 0.1 as is required for dark matter.

These hypothetical dark matter particles not only interact with one another, but they also very weakly interact with normal matter; they can, in principle, elastically scatter the nuclei of atoms. In the context of supersymmetry as it now stands, the cross sections for these interactions are essentially unknown – different models make predictions which differ by many orders of magnitude. But herein lies a possibility of direct detection of dark matter particles in terrestrial laboratories. The disk of the Milky Way Galaxy is presumably rotating through a massive dark matter halo – a sea of dark matter particles which is not rotating itself. So the Earth is moving through a wind of WIMPs – a wind with a mean velocity of about 200 km/s. In addition, the WIMPs themselves have a random velocity of the same magnitude. Now and then, one of the WIMPs will interact with an atomic nucleus and scatter that nucleus. Direct detection of the scattered nuclei is the principle behind most direct dark matter searches.

Self-annihilation of dark matter particles generally ended in the very early Universe when the density became too low for particles to encounter each other frequently. But, since that time structure has formed via gravitational collapse, and the dark matter particle density has increased again in selected regions – for example, the center of the Milky Way Galaxy or in the putative dark matter satellites of the Galaxy. Even in the Sun and Earth, some dark matter particles will scatter atomic nuclei, lose energy and become trapped. So dark matter particles should accumulate in the interiors of the Sun and Earth.

In such regions, where the dark matter particle density has increased, we would expect an enhanced self-annihilation, and it might be possible to detect the products of this process – neutrinos, gamma-rays, electrons and positrons (see Figs. 11.1 and 11.2). This is the principle underlying indirect detection of dark matter.

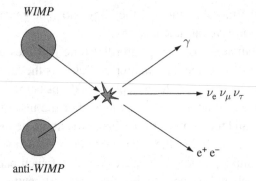

Fig. 11.1. The WIMP and its anti-particle annihilate and, this event, after intermediate steps in which unstable particles are produced, results in various standard-model end products: photons, neutrinos, and electrons and positrons. The proportions of the end products are model dependent, but they are all, in principle, experimentally detectable from discrete sources with a high density of WIMPs.

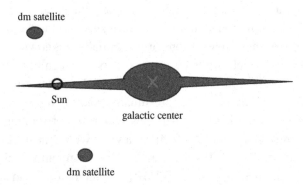

Fig. 11.2. Sources of products of WIMP annihilation would be the galactic center and nearby dark matter dominated satellite galaxies, as well as the Sun and the Earth where WIMPs are trapped and accumulate.

11.2 The practice of direct detection

In the state of Minnesota near the picturesque shores of Lake Vermilion, there is a mine shaft extending down more than 700 meters below the surface. In the nineteenth century this was an extremely productive iron ore mine – the Soudan mine – but by 1962 mining had essentially ceased, and the US Steel Corporation donated the site to the state of Minnesota which in turn designated the mine (in 1979) as an underground physics laboratory (Fig. 11.3). Such a setting is ideal for experiments which require a low background of cosmic rays as does, for example, direct detection of supersymmetric dark matter particles.

Fig. 11.3. Prospecting for dark matter at the Soudan mine, or at least the part visible from the surface of the Earth on a nice (summer) day.

In 1999 a large collaboration, 12 universities and institutions, began an experiment in the Soudan mine with the goal of directly detecting dark matter particles presumed to be a supersymmetric partner. The idea was to look for the very occasional nuclear recoil in a semiconducting material such as germanium. Basically it works like this: the hypothetical dark matter particles in the Milky Way halo have a velocity of 200 to 300 km/s. If their mass were of the order of 100 GeV this would mean that their typical energy is about 50 keV; the dark matter particles could impart this much energy to an atomic nucleus, so it is quite a low-energy experiment. The semiconducting material, silicon and germanium, is cooled to a fraction of a degree above absolute zero. On those extremely rare occasions when a passing WIMP scatters a nucleus, this will excite sound waves (the sound waves are actually quantized and are called "phonons"). The sound waves heat up the semiconductor and change its resistance to electrical current. In this way they can be detected. Because of the necessary cooling the experiment is called the "cryogenic dark matter search" or CDMS.

The biggest problem with this sort of experiment is the background. Taking a very generous value for the WIMP–nucleon scattering cross section it is expected that there will be less than one scattering event in 10 days per kilogram of germanium. In an unshielded environment the background can be many times larger – easily 100 000 times larger. Cosmic rays are a principal source of false signals and this is why the experiment is far below ground; the cosmic ray background is strongly reduced at this depth, but some cosmic rays penetrate even into this deep mine. The cosmic ray photons, gamma rays, ionize the atoms of the semiconductor and produce electrons. The electrons also excite phonons, so it is useful to have a technique for discriminating electrons from nuclear recoils. The ratio of ionization to heat seems to be an effective filter: the gamma ray events produce more ionization relative to the true nuclear recoils.

Local radioactivity is another serious background problem: there is natural radioactivity in the rock walls of the mine, and some fraction of this is in the form of neutrons. Encounters of neutrons with nuclei in the detector mimic WIMP–nucleon scattering so further steps must be taken to shield the detectors within the mine. The materials that make up the detector, the semiconductors and the electronics, can also be a source of background and must be extremely radio-pure.

The CDMS experiment has been operating for about 10 years now and has not definitively detected a WIMP. It has, however, significantly constrained the cross section of the WIMP–nuclear interaction over a wide range of WIMP masses. This is displayed in Fig. 11.4.

On this plot of cross section vs. mass, the area above the solid dark curve has been excluded (as of late 2008) with high confidence by the CDMS experiment. These limits are continually getting tighter and are beginning to constrain supersymmetric models. A number of other direct-detection experiments, using different methods for finding nuclear recoils and different materials, have also been underway – and generally report similar results (more recently the XENON100 experiment has pushed the limit on the WIMP-nucleon scattering cross-section a factor of 30 below that of CDMS, see Aprile, *et al.* 2012). There is one exception, and that is the DAMA experiment in Italy.

As the Sun orbits about the center of the Galaxy on a roughly circular orbit about once in 100 million years, it moves through the putative dark halo having a local density of about 5×10^{-24} g/cm^2 (this figure is based on disk and halo models for the galactic rotation curve). This mass density corresponds to one WIMP locally in every 35 cubic centimeters if the mass is 100 GeV. There may now be two WIMPs in your cup of coffee.

The Earth, of course, is also orbiting the Sun once a year with an orbital velocity of 30 km/s, making an angle of about 60 degrees to the direction of the Sun's motion. This means that the mean velocity of the Earth with respect to the WIMP halo varies periodically during the year; in June the speed of the WIMP wind is about 10% higher than it is in December. The higher the WIMP wind velocity, the larger the detection rate, so in principle this variation could yield an annual modulation in the WIMP detection signal of a few percent.

The DAMA experiment is designed to look for this annual modulation signal. DAMA does not use the cryogenic method – the generation of phonons in a semi-conductor – to look for nuclear recoils due to WIMP interactions, but a rather different technology: solid scintillation detectors consisting of sodium iodide. In a crystal of sodium iodide a single nuclear recoil event will produce about 40 photons and roughly 10% of these can be detected by photomultiplier tubes. There is, of course, always a cosmic ray background, so the DAMA experiment is also located in a deep tunnel under the Gran Sasso massif in central Italy. To some extent, the scintillation technique can distinguish between signal and various sources of background by the decay time of a scintillation event; for example, free electrons produced by a gamma-ray ionization of atoms have a scintillation pulse with a shorter decay time than do nuclear recoils, although there is overlap at the keV energies of interest.

DAMA, in a sense, sidesteps the problem of background by looking for an annual modulation in the entire signal. The results over about 12 years are shown here in Fig. 11.5. The data is clearly in two sequences with the final three years showing a significantly higher signal-to-noise. This was due to an upgrade of the detectors (a larger mass of sodium iodide) in 2003. An annual modulation is clearly present with an amplitude of about 2% in the energy range of 2 to 5 keV. The modulation is evident only at lower energies as would be expected in the scattering of nuclei by galactic WIMPs. Strikingly, the phase of the modulation is also as it should be for the WIMP wind: the maximum occurs in June and the minimum in December.

Has DAMA detected the WIMP dark matter? DAMA clearly has seen an annual modulation of low-energy events of about the right phase, but the dark matter interpretation remains controversial, basically because a signal of this strength should have been seen in other more sensitive experiments. For example, the CDMS experiment claims a sensitivity much greater than that of DAMA; and the DAMA result lies well into the excluded region of the cross section vs. mass plot (Fig. 11.4). It is apparently ruled out by the most recent CDMS results. Until recently there has been one escape for DAMA: the WIMP–nucleon interaction may be dependent upon the spin of the nucleus. Germanium (the detector material in CDMS) has an even number of protons (atomic number 32) and exists in several isotopes with varying numbers of neutrons; the net nuclear spin is near zero. This means that CDMS is sensitive to spin-independent interactions. But the nuclei of sodium and iodine (the stuff of DAMA) both have odd numbers of protons and so a net spin. If the WIMP interacts more strongly with nuclei having a net spin, then this could explain why DAMA sees them but CDMS does not.

A recent experiment at the University of Chicago Fermi Lab seems to close this loophole (COUPP, "Chicagoland Observatory for Underground Particle Physics").

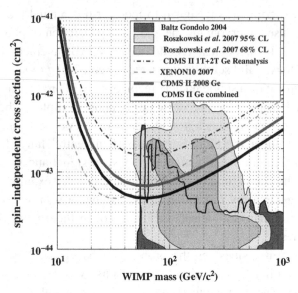

Fig. 11.4. This is a plot of cross section vs. mass of hypothetical dark matter particles. The solid black line shows the limits set by the current data on nuclear recoils by the CDMS experiment in the Soudan mine. The area above this line is ruled out by this experiment with a confidence of 90%. The area below the curve is still permitted; that is to say, the WIMPs may have properties below this line. For example, if the WIMP mass were 100 GeV, then the cross section could be as large as 4×10^{-44} cm^2. The shaded regions in the lower right show the predicted ranges of mass and cross section for various supersymmetric models. This data is already beginning to seriously constrain such models. From Ahmed *et al.* (2009).

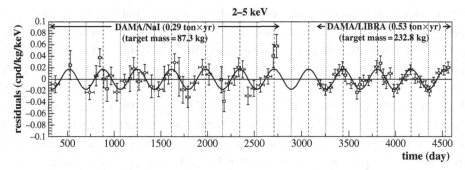

Fig. 11.5. Residuals about the mean signal with DAMA. Higher signal-to-noise data with the upgraded system is evident. From Bernabei *et al.* (2008).

This experiment uses a quite old technology – the bubble chamber. The bubble chamber contains a superheated liquid; i.e., a liquid at a temperature higher than its boiling point. In this case the liquid is trifluoroiodomethane (CF$_3$I). This molecule is about 30% fluorine which has nine protons and therefore a net nuclear spin; it should be very sensitive to spin-dependent WIMP interactions. When a nucleus struck by a WIMP passes through the superheated CF_3I the resultant heat nucleates bubbles of vapor phase which then trace the path of the particle. This experiment could detect WIMP–nucleon scattering as well as scattering due to random background neutrons, although an advantage is that the apparatus is self-shielding for neutrons – neutrons cannot penetrate far within the liquid.

The initial results from this experiment find no spin-dependent WIMP–nucleon scatterings at a level that is inconsistent with the DAMA result which would seem to seal this particular escape hatch for DAMA (Behnke *et al.* 2008). On the other hand, the DAMA group argues that their experiment is sensitive to a range of dark matter candidates that do not appear in experiments looking for recoiling nuclei. The fair conclusion is that the DAMA result remains controversial. Until the claimed detection is confirmed by another experiment – hopefully using a different technique – then one cannot argue that the dark matter particles have been detected. We must be aware that DAMA is looking at a modulation of the entire signal including the background, so it cannot be excluded that the background itself has an annual modulation. A number of physical effects do vary on the timescale of one year: temperature, background radon concentration (a radioactive gas producing ionizing radiation), electricity usage on the national grid. One of these effects may or may not be the culprit but these possibilities do stress that the result must be confirmed by another experiment before it is credible.

Finally, when considering direct detection, we should keep in mind that the LSP or a higher-mass superpartner may be directly created in large high-energy particle accelerators. The "large hadron collider" (LHC) has recently been completed at the European Council for Nuclear Research in Geneva (CERN), and it will accelerate and collide protons up to energies of several TeV (1000 GeV). The accelerator is now the world's largest, operating at the highest energy, and is expected to be fully operational in 2010. If supersymmetric particles have masses of 10 to 100 GeV, the LHC may produce them in high-energy collisions; these particles would be apparent as missing energy – energy not accounted for in the collision products. If these events are seen, it should be possible to determine certain properties of the WIMPs, such as the mass, but other relevant properties, such as the annihilation cross section are model dependent. This means that even if the lightest superpartner is detected at the LHC, it may not be the dark matter. Supersymmetry may be an appropriate extension of the standard model, and supersymmetric partners may exist but still not have the correct properties to constitute the dark matter; the relic abundance may be far too low. In any case, this, and other direct-detection experiments will provide tighter and tighter constraints on the possibility of WIMPs as dark matter.

There are numerous experiments, not mentioned here, using different detection materials and technologies. This is important because any claimed detection must be confirmed preferably by other methods. The reader has probably also noticed that no names are mentioned here. It is difficult to attribute credit to individuals because of the large number of people involved in dark matter searches; for example, the most recent CDMS paper has 58 co-authors. Several individuals do stand out: Bernard Sadoulet (2007) at Berkeley is a pioneer in cryogenic dark matter detection and has been an active advocate for such experiments for many years. A classic paper, a kind of Bible on the theory and strategy for direct and indirect detection of supersymmetric dark matter particles, was published in 1996 by Gerard Jungman (Syracuse), Marc Kamionkowski (Columbia) and Kim Greist (UC San Diego). Richard Gaitskell of Brown University is a member of the CDMS team (with many others), and has written an important review on the subject of direct dark matter detection (2004). Rita Bernabei is the leader of the DAMA experiment and has forcefully defended this controversial result (2008).

With respect to direct detection, I should mention one more possibility. It may be that the dark matter does not consist of WIMPs; the particles would certainly have to be weakly interacting, but they may not be massive. The "axion" is such a theoretical particle proposed to solve a specific problem with the standard model. The theory of the strong force, the force that confines the quarks within a proton or neutron, is called quantum chromodynamics. This theory in its original form permits the violation of a certain kind of symmetry called "charge-parity" (CP). A consequence of this symmetry violation is that the neutron should possess an electric dipole moment, even though it is electrically neutral. No such dipole moment is measured, which implies that the strong interaction does in fact respect this CP symmetry. This can be accomplished if a new element is added to the theory: a very low-mass particle that has been called the axion. Even though the axion has a very low-mass, it is born cold; it was never in thermal equilibrium with the cosmic fluid (this is because the axion may be represented as an oscillation of a field in a certain kind of potential; the oscillations damp rapidly as the axion acquires its mass in the early Universe). Thus, the axion is considered to be a well-motivated candidate for cold dark matter.

These hypothetical particles interact with a magnetic field; they scatter the virtual photons of the field and produce real photons in the frequency range 100 MHz to 100 GHz (radio waves) depending on their mass. So the search strategy is to look for the photons produced by scattering of galactic halo axions in the presence of a strong magnetic field within a finely tuned radio-frequency cavity at a low temperature. One such search is the ADMX project (axion dark matter experiment) at Lawrence Livermore National Laboratory. So far nothing has been seen (after an earlier claimed false detection by a different group); the mass of this particle is now constrained to be less than 2μ eV (2×10^{-6} eV) for a particular class of axions (Duffy *et al.* 2006). This work continues with higher sensitivity.

11.3 Indirect detection of dark matter

As the Sun with its planets moves through the galactic sea of dark matter, WIMPs will encounter the Sun and the Earth. Because they are, after all, weakly interacting, most will pass through with their paths deflected somewhat by the Sun's gravitational force. But some of the particles will scatter nucleons and lose energy. If they lose enough energy they will be captured by the Sun and by the Earth. The concentration of WIMPs in the Sun (or the Earth) increases to the point where the WIMP–WIMP annihilation rate equals the capture rate. The products of WIMP annihilation are standard-model particles: pions, muons, photons, electrons and positrons and neutrinos (Fig. 11.1). Therein lies a possibility of detection of WIMPs. Because the WIMPs, or their interaction with matter, is not being detected directly, this is called "indirect detection"; the decay products are detected.

In the Sun, the decays are occurring deep in the solar interior; the only decay product that can stream directly from the interior to detectors on the Earth are neutrinos (recall neutrinos have a very small cross section for interaction with matter). Therefore, dark matter decays (WIMP burning) occurring in the Sun or Earth are possibly detectable by the resulting neutrinos. These, of course, would be extremely high-energy neutrinos, with energies going up to the rest-mass energy of the WIMP, perhaps several hundred GeV.

Neutrino astronomy represents a very new window on the Universe, and there are now several "neutrino telescopes" which are in operation or planned. Neutrinos do interact occasionally with atomic nuclei producing other subatomic particles called muons. The muons produced by energetic neutrinos are moving near the speed of light – in fact, if the interaction takes place in a medium such as water, the muons are moving faster than the speed of light. In this case they emit electromagnetic radiation called Cherenkov radiation in a narrow cone about the direction of motion, and the Cherenkov radiation may be detected by photomultiplier tubes. With a large three-dimensional array of photomultiplier tubes the path of the muon, and hence of the original neutrino, may be deduced.

A neutrino telescope consists of an array of photomultipliers suspended in a medium such as water or ice. As always there is a background of muons produced by atmospheric cosmic rays, but these muons are traveling downward. So the neutrino telescope uses the entire Earth as a shield. The cosmic ray muons cannot penetrate the Earth, but the muons of interest are produced by neutrinos that can easily penetrate the entire Earth. Unlike the usual astronomical telescopes, neutrino telescopes do not look up – they look down; the events of interest are those coming up from the other side of the Earth. WIMPs are not only captured by the Sun, but they would be expected to accumulate in the Earth by the same mechanism. The neutrino telescopes might also be able to detect neutrinos from decaying WIMPs in the core of the Earth.

One such neutrino telescope that is already operational is the ANTARES (astronomy with a neutrino telescope and abyss environmental research) telescope in the Mediterranean sea off the coast of southern France. An artist's impression is shown in Fig. 11.6. Such large projects have primary scientific goals other than the possibility of indirect dark matter detection; explosive objects, such as supernovae and possibly gamma-ray burst sources, also emit neutrinos, and the goal is to observe the neutrino emission from these objects. But very energetic neutrinos, in excess 10 GeV, arriving from the direction of the Sun (observed, of course, at night) or from the very center of the Earth would constitute evidence for dark matter annihilation.

Other instruments that are operational or in construction are AMANDA and its extension, ICECUBE. For these neutrino telescopes, photomultiplier arrays are buried deep in the ice shelf on the Antarctic continent. Ice, as a medium, has a lower natural background than sea water, and the ice in Antarctica is very transparent.

Fig. 11.6. An artistic view of the undersea ANTARES neutrino telescope. The three-dimensional array is comprised of 12 vertical strings of photomultiplier tubes anchored to the bottom of the Mediterranean Sea at a depth of 2.5 km. Courtesy of the ANTARES collaboration.

AMANDA is already placing constraints on the rate of dark matter decay in the center of the Earth. ICECUBE, currently being constructed, is a truly mammoth project with 70 one-kilometer strings of 60 photomultipliers extending down about 2 km into the Antarctic ice. With a total volume of one cubic kilometer the array will provide high sensitivity and resolution for energetic neutrinos. The project will address fundamental questions about the most energetic sources in the Universe independently of the dark matter problem. This kind of project, that does not just pursue a single scientific goal – particularly one that is dependent upon the unknown detailed properties of a hypothetical dark matter particle – is clearly worth the large investment of money and effort.

With respect to dark matter, keeping in mind that neutrino telescopes are only now (2009) becoming fully operational, it can be said that there is no evidence for neutrinos from WIMP decay. However, there are additional products of WIMP decay that may be accessible to other instruments. For example, one might expect to observe very high-energy photons, gamma rays, and, perhaps even an annihilation line – photons at a particular frequency corresponding to the rest-mass of the WIMP. This, in a sense would be the "smoking gun" of indirect dark matter detection because it is difficult to conceive of another astrophysical mechanism that could produce such a signature.

Several instruments are capable of addressing this issue now. The "Fermi gamma-ray space telescope" (FGST, formerly GLAST; see fermi.gcfc.nasa.gov/public), launched in 2008, has the possibility of detecting photons emitted from decaying dark matter particles in environments where the density of dark matter is presumably high, such as the galactic center or dwarf satellite galaxies in the galactic halo (Fig. 11.2). FGST can detect photons in the relevant energy range (up to several hundred GeV) and might see annihilation lines. In this regard it is interesting to point out a possible connection with an unexplained phenomenon seen by WMAP; the CMB anisotropy experiment. WMAP has not only observed the anisotropies in the CMB but has also made a map of the galaxy at microwave frequencies. In an extended region about the galactic center a diffuse emission is seen at a wavelength of about 1 cm. Several people such as Dan Hooper at Fermi Labs have proposed that this WMAP "haze" is due to high-energy electrons and positrons produced by decaying dark matter. These charged particles spiral in the galactic gravitational field and produce radiation, synchrotron radiation, which is observable at radio frequencies. If this is the explanation for the WMAP haze then gamma-ray lines should be visible with FGST provided that the decaying dark matter particles have a mass less than a few hundred GeV.

Several ground based gamma-ray telescopes are currently operational and may detect decaying dark matter from discrete sources. Very high-energy gamma rays

Fig. 11.7. The HESS gamma-ray telescope is located in the Khomas highlands of Namibia, a very dry region with excellent conditions for observing in the optical. Each of the 13-m telescopes consists of a collection of mirrors which bring the Cherenkov light to the focus. The multiple-telescope array allows the position of the gamma-ray source to be determined. Courtesy of the HESS collaboration.

cannot penetrate to the surface of the Earth (fortunately). When a gamma ray, for example, a photon with an energy of 100 GeV, enters the atmosphere of the Earth, it interacts with the atmospheric nuclei and initiates an electromagnetic air shower. That is to say, the gamma ray produces high-energy particles which decay into electrons and positrons. These particles can be detected because they are moving faster than the speed of light in air and, so, emit Cherenkov radiation. The ground-based gamma-ray detectors are actually optical telescopes, usually more than one for accurate positioning, that look for this characteristic Cherenkov radiation. Two such telescopes in operation now are VERITAS (the "very energetic radiation-imaging telescope array system") located on Mt. Hopkins in the state of Arizona, and HESS (the "high-energy stereoscopic system") in the dry desert of Namibia (Ahronian *et al.* 1997, see Fig. 11.7). These instruments are capable of detecting gamma rays in the energy range of 50 to 50 000 GeV (relevant for dark matter decay), and they also can determine the position on the sky of the source of gamma radiation. Pointed observations of nearby presumed concentrations of dark matter, dwarf spheroidal galaxies, have so far found no evidence for dark matter decay.

The EGRET detector (the "energetic gamma-ray experiment telescope") on board the Compton satellite has detected a source of diffuse gamma radiation with energies from 100 MeV to 10 GeV in an extended region about the Galactic center. The flux of this source exceeds that which is expected due to the interaction of cosmic rays with the interstellar medium (the diffuse gas, mostly hydrogen, between the stars). It has been suggested that this source could result from dark matter decay, but there are alternative astrophysical possibilities (scattering of starlight and CMB photons up to a high energy by cosmic rays).

Therein lies the essential problem with indirect dark matter detection. Whenever an unexpected signal is detected, it is tempting to attribute this to dark matter decay, but often an alternative physical process may be responsible. It is difficult to find the smoking gun.

The decay of dark matter particles not only produces gamma rays; depending upon the kind of dark matter and upon the decay mode, this process may also produce primarily particles: protons, anti-protons, electrons and positrons. So detectors which can determine the energy distribution of the electron and positron components of the cosmic rays are also relevant to this problem.

There are two recent observations that have produced provocative results in this respect (provocative in the sense that many interpretive papers have appeared in the recent literature, even before publication of the observational results). The PAMELA satellite ("payload for antimatter/matter exploration and light-nuclei astrophysics") was launched in 2006 by a European consortium. This satellite is capable of measuring the cosmic-ray electron and positron energy distributions separately up to energies of about 80 GeV. There are primary cosmic-ray electrons, but positrons and electrons are also produced as secondary particles when heavier cosmic rays (protons for example) interact with the interstellar medium. A general calculation, which takes into account both the production of positrons and electrons by these processes, as well as the propagation of particles through the Galaxy, predicts that the ratio of positrons to electrons should decrease with energy. But, in fact, above energies of about 10 GeV, the ratio actually increases. This implies that there should be another source of energetic positrons in addition to cosmic-ray production of secondary particles. Dark matter decay is one possibility, but the source must be rather close (a nearby dwarf dark halo, for example). High-energy electrons and positrons do not travel far in the Galaxy because of energy loss due to synchrotron radiation, the spiraling of the particles in the interstellar magnetic field. Another possibility is a different sort of WIMP which decays preferentially into electrons and positrons.

A second experiment is the ATIC ("advanced thin-ionization calorimeter"). This is a balloon observation (also launched in Antarctica) of the electron and positron cosmic-ray spectrum, but it cannot distinguish between electrons and positrons. ATIC sees an unexpected increase in the total electron–positron flux up to about 620 GeV and then an abrupt decrease (see Fig. 11.8). This is the sort of feature that

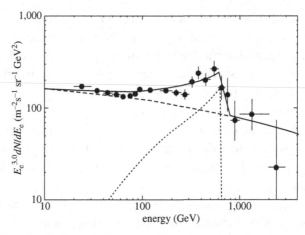

Fig. 11.8. The distribution of electron and positron energies observed by ATIC. The upturn and feature at about 600 GeV is clearly visible. Note that the energy distribution is multiplied by E^3 to remove the general power-law trend and enhance the appearance of the feature. The dotted curve is the energy distribution predicted by a particular dark matter annihilation model. Although this is the characteristic signal expected from dark matter annihilation, FERMI and HESS fail to confirm this feature. From Chang *et al.* (2008).

one might expect from a decaying dark matter particle of 620 GeV, but again the particle would not be a standard supersymmetric partner but something rather more exotic. Moreover, the source must be quite close, within 1 kpc, or the 620 GeV electrons/positrons could not reach us due to the energy losses.

Two recent papers (May 2009) by the FGST and HESS collaborations (more than 100 authors on each paper) report results relevant to this claimed feature. Both instruments, designed to detect high-energy gamma rays, can also detect high-energy electrons and positrons (but also cannot discriminate between them). For example, HESS detects air showers due to pair creation by gamma rays, but it can also see Cherenkov radiation due to showers from primary electrons and positrons themselves. With larger sensitivities than ATIC, FGST and HESS do not see the feature reported by ATIC. This shows the necessity of confirming any reported signal with other detectors as well as the risks of jumping too soon onto an interpretive bandwagon; this field is fiercely developing.

In any case, both PAMELA and ATIC results have more standard astrophysical explanations, for example, spinning neutron stars (pulsars). These dense rapidly rotating compact objects are expected to produce energetic electrons and positrons near their surface where the magnetic field is extremely high. One or several nearby pulsars might account for local electron and positron excesses.

11.4 Light on dark matter: the story so far

It is clear that a large number of physicists are willing to devote their careers to the direct detection of dark matter particles – this, in spite of the uncertain nature of the particles and their properties, not to mention the fact that even the existence of such particles is hypothetical. Given the range of possible cross sections, non-detection in experiments currently underway or planned will not rule out WIMPs as the dark matter. Moreover, the ingenuity and imagination of theoretical physicists can always accommodate any non-detection by inventing new possible dark matter candidates. The point is – non-detection is not falsification.

Nonetheless, the intense investment of time, energy and money in direct-detection experiments reflects the general perception that the problem is of over-whelming importance. After all, what could be more significant than the identity of 80% of the matter content of the Universe? The rewards of direct detection would be considerable. There is also the enormous appeal of doing fundamental research in particle physics without going to extreme energies; it is remarkable that the very nature of matter and of the underlying symmetries of nature can be investigated using a bubble chamber.

So far, in the experiments currently underway, there is no convincing evidence for the direct detection of dark matter particles. The DAMA result is provocative but controversial in view of non-detection in experiments claiming to be more sensitive. This is clearly a controversy which should be resolved. The sensitivity of detectors and the variety of technologies is increasing rapidly. But if there is no detection in the following decade, it will be interesting to see how many physicists are still involved in this line of research.

Indirect detection has also yielded no definitive results so far, although the field is developing rapidly (dangerous for a book given the delay time between writing and publishing). Also here, there are suggestive hints (such as the positron excess), but most of these require rather special properties of the dark matter particles or special locations of the source of decays. And again, the problem with indirect detection is that alternative astrophysical explanations are generally available. It is probably a mistake to assume that any unexpected signal is due to dark matter decay. It would seem to be a general requirement that any indirect signal of dark matter must be confirmed by direct detection before breaking out the champagne (or buying tickets to Stockholm).

12

Reflections: a personal point of view

The prevailing view of the Universe now is radically different than it was 40 years ago when I began my career as a professional astronomer. Then, the world was perceived to be essentially visible to our eyes and our instruments. Now less than 5% of the world is thought to be potentially visible, and even 90% of this normal baryonic component is not detected. The remainder of the mass–energy content of the Universe is thought to consist partly of dark matter that is unidentified, and primarily of dark energy of even more uncertain nature. The dark matter fills the Universe, promotes structure formation and accounts for the discrepancy between the visible and dynamical mass of bound astronomical systems such as galaxies and clusters; it is the major constituent of such systems. In order to cluster on the scale of galaxies at sufficiently early epochs, the dark matter must be essentially pressureless, i.e., non-relativistic at the time it decoupled from photons and other particles. The dark energy, which may be identified with the zero-point energy of the vacuum, causes the present accelerated expansion of the Universe and provides the 70% contribution to the density budget for the flat Universe required by observations of CMB anisotropies.

So the two primary constituents of the world are detected only by their dynamical effects: dark energy which affects in an observable way the expansion history of the Universe, and dark matter that clumps and dominates the mass budget of bound gravitating systems. Some members of our scientific community are uncomfortable with the concept of dark energy. There is a theoretical problem with its small magnitude (on the scale of particle physics). Moreover, in so far as this peculiar fluid can be represented by a cosmological constant, it does not dilute with the expansion of the Universe as does the density of ordinary matter. Why, then, are the densities of these two substances so nearly comparable at the present epoch? Why are we now witness to this remarkable coincidence? If the dark energy is dynamical and identified with some new field, should we not see other manifestations of that field, such as fifth-force effects, evident as a violation of the universality of free-fall?

People are more comfortable with the concept of dark matter because it is more comprehensible. Dark matter is appealing to two scientific communities: physicists and astronomers. Theoretical physicists like it because the best-bet extension of their standard model, supersymmetry, provides candidate particles with, possibly, the right properties. Many very competent experimentalists are willing to spend a significant fraction of their lives looking for it in the laboratory, even though its nature, and therefore the detection strategy, remains uncertain. The search for dark matter has become a large industry with all the vested interests of a large industry; this is because detection would be one of the major scientific discoveries of all time. But I have tried to make the point here that non-detection is not falsification, even in the context of a particular class of dark matter candidates; superpartners for example. And given that the range of possible candidates is limited only by the human imagination, then unsuccessful dark matter searches can always be accommodated in the context of the paradigm.

Astronomers like dark matter. If you cannot see it, you can use it to produce rotation curves of any sort, stabilize disks, make warps, promote mergers, explain anomalous lensing, make large-scale structure... It is fun to simulate because all you need is Newtonian gravity which is easy to compute. Again, the one thing you cannot do is falsify it; almost any astronomical observation can be accommodated by dark matter. Systematics of galaxy rotation curves and near-perfect global scaling relations can be ignored because these result from poorly understood "gastrophysics" – gas dynamical process, star formation, feedback. This of course is accompanied by a large leap of faith that someday these processes will be understood and all will be explained.

While these two communities, physicists and astronomers, have found a common interest in dark matter, we should remember that they are, in fact, rather distinct communities with different methodologies and different criteria for the interpretation of results. Physicists are more prepared to go beyond known physics (supersymmetry, after all, is an extension of known physics) while astronomers are more conservative in this respect (and well they should be; interpretation of astronomical results can otherwise become quite bizarre). Physicists know that galaxy rotation curves are flat and that this constitutes a primary evidence for dark matter that should manifest itself locally. They do not know about, and are not much interested in, the regularities of rotation curves or global scaling relations; these are details for astronomers. Astronomers know, because physicists tell them, that particle dark matter is well motivated and that it is proper to invoke dark matter in understanding astronomical observations.

Of course, this division between physicists and astronomers is simplistic; there are individuals who move easily between both communities, have a broader overview, and strongly support the concept of dark matter. The point is that both physicists and astronomers find it useful to invoke dark matter from their own different vantage points, but I argue that from both sides the dark matter concept is fundamentally not falsifiable.

It was Karl Popper who first emphasized the importance of falsification in eliminating scientific theories and progressing to new ideas. This surely must be true, given the inherent asymmetry between falsification and verification. Of course, to be falsified a theory must be in practice falsifiable; this would seem to be a hallmark of good theory (for Popper, a theory that is not falsifiable is not scientific). Dark matter as a theory misses this attribute (which is not to say that it is wrong). In my opinion the most serious challenge for the dark matter hypothesis is the existence of an algorithm – MOND – that can predict the form of rotation curves from the observed distribution of detectable matter. This is something that dark matter does not naturally permit because it is a different sort of fluid and not subject to all of the physical effects that influence baryonic matter and its distribution. Moreover, MOND, as a theory, is inherently falsifiable. If particles with the right properties to constitute the cold dark matter are found tomorrow, then MOND is out of the window. In that sense it is a better theory (which is not to say that it is right).

With respect to progress through falsification, the reality is never so simple and certainly is not in this case. In the issue of dark matter vs. MOND we are not dealing just with two theories but with two competing paradigms in the sense meant by Thomas Kuhn. Thirty years ago it was becoming generally recognized that something was missing in large astronomical systems like galaxies and clusters. This recognition was not an instantaneous process as I have discussed in Chapter 4. When an observation runs counter to our expectation, we do not always perceive the anomaly; recall the early attempts to fit observed rotation curves exhibiting no evidence for a decreasing rotation velocity by models having a built-in Keplerian decline. But by 1980, it was no longer in doubt that *something* had been discovered; but *what* that something is, in fact, has never been so certain. Dark matter was the initial, and natural, first attempt at a solution to this astronomical anomaly. With respect to galaxies, the concept of dark halos was already in place as a means of taming the instability of rotationally supported systems. At the same time, it became appreciated that the difficulty of forming the observed structure in an expanding Universe with a finite lifetime could be overcome by adding a universal non-baryonic matter component. And at the same time, particle physics seemed to be providing a host of particle candidates. The astronomical anomaly, the cosmological necessity, and the particle physics possibility combined to give the dark matter hypothesis the status of a paradigm: a framework, a set of assumptions that are not questioned, a list of problems that are to be addressed as well as problems that are not to be addressed.

Not long after the discovery of the anomaly (1983), the hypothesis of modified Newtonian dynamics (MOND) emerged, and this proposal can be clearly associated with a single individual – Milgrom (there were other such ideas in circulation, but none of them successfully addressed so many aspects of the phenomena). At the time MOND was what Kuhn would call an "anticipation" and not a response to a crisis with dark matter – there was no such crisis. MOND was truly an alternative to the dark matter hypothesis – in fact, the only alternative explanation for the observed anomaly. But while the dark matter hypothesis attracted a large following early on – thanks primarily to its range of application, from galaxies to cosmology – MOND languished for some years with only a handful of advocates (myself included). But now due to its proven predictive power, at least on the scale of galaxies, the development of a reasonable relativistic extension (thanks primarily to Bekenstein), and simple frustration with the absence of dark matter particle detection, MOND has also achieved the status of a competing paradigm, although one still supported by a small minority of the relevant communities.

Supporters of different paradigms give different weight to different experimental or observational facts, and this makes the issue of falsification rather murky. For example, supporters of the dark matter paradigm tend to emphasize cosmological aspects. They would argue that on a cosmological scale general relativity with dark matter (and dark energy) presents a coherent picture. The observed phenomenology of the CMB fluctuations and the formation and distribution of galaxies on a large scale is explained in the context of the concordance cosmology. They tend to dismiss galaxy-scale phenomenology and its systematics as being essentially due to messy baryonic physics which will someday be understood in the context of the larger picture. MOND supporters, on the other hand, emphasize the regularities in galaxy phenomena: the predicted appearance of a discrepancy in low-surface-brightness systems, the near perfect Tully–Fisher law, the ability of the algorithm to predict the amplitude and form of rotation curves. They are rather dismissive of the cosmological evidence, at least until the recent development of the relativistic extension.

The point is (and this is essentially Kuhn's point) arguments between supporters of different paradigms are somewhat akin to arguments about religion. The assumptions and the criteria for truth are different. Most scientists do not feel the need to adopt a new paradigm unless the old one is in crisis, so we may ask: is dark matter in crisis? Are there fundamentally un-resolvable anomalies within the context of dark matter? Again, most supporters would answer in the negative (but they would then, wouldn't they?). I personally think that there is a crisis – more of a creeping crisis provoked by the non-detection of dark matter particles. I have argued that this is not properly a falsification, but it surely must be a worry. At what point will experimentalists stop searching for these elusive particles and shift to activities more likely to produce positive results? At what point will theorists tire of more and more speculative conjectures on the nature of hypothetical undetectable matter? And what if apparent deviations from Newtonian gravity or dynamics are seen in the Solar System? What if the Pioneer anomaly is confirmed and clarified?

It is certainly true that, for scientists taken as a social group, most effort goes into attempting to prove or strengthen the existing paradigm rather than to challenge it. Consistencies are valued over anomalies, and uncomfortable facts are overlooked or pushed into the category of complicated problems for the future. This is probably necessary because normal science takes place in the context of a paradigm. The social phenomenon is reinforced by external considerations: by competition for academic positions, by the necessity of obtaining research grants. I expect that it has always been this way, but in a general sense (and this is why Kuhn emphasizes the significance of "scientific revolutions") progress is a dialectic process and due to the conflict of ideas rather than "concordance". By "progress" here I mean moving in a direction of increased understanding of the world around us. Kuhn would certainly not agree with this definition, nor even with the concept of progress as movement toward a goal, but I believe that it is meaningful.

With respect to MOND, I have been impressed because it explains and unifies aspects of galaxy phenomenology which would appear to be disconnected in the context of dark matter. Viewed in the context of MOND, the "scales have fallen from my eyes". For me, the question is not – Is it right? – but – Is it timely? In this sense it may be relevant to recall Alfred Wegener's original proposal of continental drift. Wegener was a meteorologist (with a PhD in astronomy) and not a geologist, so he was an outsider to that particular community. But when he presented this idea in 1912 he demonstrated that the concept explained a number of apparently unrelated facts, in paleontology as well as geology – not just the fact that continents appeared to fit together like pieces of a puzzle. For example, the idea accounted for the similarity of fossilized species on the west coast of Africa to those on the east coast of South America. It explained the evidence for a past tropical climate in Antarctica. It goes without saying that his idea was met with ridicule by the geological establishment. This was not because all geologists were close-minded, blind fools (although some of them undoubtedly were); it was rather because no one, including Wegener himself, could conceive of a mechanism by which the massive continents could drift across the floor of oceans. It was not until 50 years later and the development of the modern theory of plate tectonics that the idea became the central paradigm of geology and recognized as the primary mechanism that structures the surface of the Earth (not that it did Wegener much good; he froze to death in 1930 on a meteorological expedition to Greenland). I do not wish to draw too close an analogy with MOND, but this does demonstrate that a proposal can be essentially correct and yet seriously premature because the underlying conceptual basis for the idea is not yet in place.

With respect to the dark matter–dark energy paradigm, in spite of all of the triumphalism about precision cosmology, it is the peculiar composition of the concordance model that calls the underlying assumptions into question, not the precision by which the model parameters are determined. There is an increasing recognition that the theory of gravity may need to be enlarged to include phenomena on a cosmic scale. This has given rise to consideration of "infrared modifications" of gravity – modifications on a large scale or low energies – such as world models in which the apparent four-dimensional space–time is a surface in a higher-dimensional space. Leakage of gravitons – the particles that carry the gravitational force – into the higher-dimensional world cause the accelerated expansion of the Universe and not dark energy per se. Higher-dimensional theories can also lead to more exotic kinds of dark matter – so-called Kaluza–Klein particles – which possess rather different properties than the standard supersymmetric WIMPs. There are also suggestions that modified gravity theories which give MOND-like phenomenology in galaxies may yield dark energy on cosmic scales or even dark matter which does not cluster on the scale of galaxies. All of this tells us that the story is far from over and that 40 years from now, our view of the Universe is likely to be quite different again than it is at present.

What has the dark matter problem taught me, in the past 40 years, about the practice of science? I think that the central lesson is that science is essentially a social activity. This may seem obvious or trivial, but for me, it has been a profound, although gradual, realization. I suppose, all those years ago at Princeton, I thought that we were in pursuit of Truth and that having glimpsed it we could return to the cave and be happy in ourselves with this very personal achievement. But only the most solipsistic individual could be satisfied with such a reward. We work essentially for the approval of a small and select community of individuals involved in similar pursuits. This community has its procedure for the training of new members, its rites of initiation, its standards of excellence, its criteria for truth. The existence of scientific communities or sub-communities is necessary for what I have called progress; it can be no other way.

The community is, in some sense, like an extended family. Perhaps this is particularly true in Holland where the community is so compact (our real families do not always understand this attachment to and dependence upon the professional family). In astronomy the family has always included a range of the usual familial types: the stately grandfather, the wise aunt, the peculiar uncle, the precocious child, the companionable cousin. There is the occasional quarrel, sometimes bitter, sometimes destructive – but in general there is a tolerance for diversity that I have always found refreshing and, finally, productive. Of course, it is a tolerance that has its limits; to disobey the basic standards and values of the family risks ejection. But I do hope that the diversity continues.

Appendix

Astronomy made simple

A1 Electromagnetic radiation

Before considering dark matter it is useful to have some understanding of visible matter. The visible matter in the Universe is primarily evident as stars. Stars are massive self-gravitating spheres of gas ($\approx 10^{30}$ kg) that are hot enough and dense enough in their interiors to produce energy by nuclear fusion – primarily the conversion of hydrogen into helium. This energy emerges from the star as electromagnetic radiation along with a flux of subatomic particles known as neutrinos.

Almost all of our knowledge of astronomical objects is derived from electromagnetic radiation. Electromagnetic radiation is a wave-like oscillating pattern of electric and magnetic fields propagating through empty space at a fixed speed of 300 000 km/s. The radiation is classified by humans according to the length of the wave (from crest to crest): radio waves have wavelengths ranging from millimeters to meters, infrared radiation, of the order of a few micrometers (10^{-6} meters), and visible light, which we can detect with our eyes, a few tenths of a micrometer (thousands of angstroms). At shorter wavelengths there are ultraviolet radiation, X-rays and γ rays.

A wave is characterized not only by its wavelength but also its frequency. This is the number of times per second a wave crest passes by an observer. Frequency, ν, and wavelength, λ, are related as

$$\lambda \nu = c \tag{A1.1}$$

where c is the speed of the wave, in this case the speed of light. Putting in numbers, this means that electromagnetic radiation in the radio regime, with a wavelength of 20 cm for example, will vibrate the electrons in a receiver about 1.4 billion times a second – a frequency of 1400 megahertz (MHz) or 1.4 gigahertz (GHz). For red light, with a wavelength of 6000 angstroms, 5×10^{14} wave crests pass into the pupil of our eye every second.

One of the two most important theoretical developments in physics in the twentieth century was the theory of quantum mechanics (the second being relativity). In the first half of the century it was realized that matter has wave-like properties, but, also, electromagnetic radiation has matter-like properties. Under some circumstances radiation appears to come in distinct bundles like particles; these particles of radiation are called "photons". Photons have an energy which is related to their frequency:

$$E = h\nu \qquad\qquad\qquad (A1.2)$$

where h, Planck's constant ($h = 6.625 \times 10^{-27}$ ergs/s), is the fundamental constant of quantum mechanics (in the limit where h approaches zero, physics becomes entirely classical). This formula would mean that photons of red light have an energy of about 2 electron volts; an electron volt (abbreviated eV) is the energy of an electron accelerated through a voltage difference of one volt. X-ray photons, with frequencies in excess of 10^{17} Hz, would have energies in excess of 1000 electron volts – a kilo-electron volt (keV). Cosmic gamma rays, with frequencies in excess of 10^{20} Hz, have energies exceeding one million electron volts or 1 MeV. In Chapter 11 we will see that gamma rays with energies exceeding one billion electron volts (1 giga-electron volts or 1 GeV) are now being observed from celestial sources.

Stars, and therefore galaxies which are built out of stars, emit radiation at all wavelengths, so-called continuum radiation. The total power of electromagnetic radiation emitted by an astronomical object is called its luminosity and this is measured as a unit of energy per unit time – such as ergs/s. The apparent brightness of an object depends not only upon its luminosity but also its distance, and physical units would be those of flux: ergs/s/cm^2. Optical astronomers generally do not use physical units but a relative logarithmic scale – the scale of apparent magnitudes. This system is ancient and due to the fact that the response of the eye to light is not linear but logarithmic. A bright star, a first-magnitude star, is 2.5 times brighter than a second-magnitude star, which is 2.5 times brighter than a third-magnitude star and so on. The ratio of the flux of two objects F_1/F_2 is related to the difference in magnitudes as

$$m_2 - m_1 = -2.5 \log(F_2/F_1). \qquad\qquad (A1.3)$$

Notice that a smaller magnitude means a larger flux. The zero point of this magnitude scale is set by some standard star such as Vega. The faintest stars we can see with the unaided eye have an apparent magnitude of about 5. Distant galaxies have a magnitude in excess of 20.

The magnitude of an object is typically measured over a specific wavelength range or color band, such as blue (B), visual (V) or infrared (K). This is made more confusing by the fact that there are several competing systems of filters covering different wavelength ranges, and conversion between them is not always straightforward.

Stars range in surface temperature between a thousand degrees (kelvin) and tens of thousands of degrees. Very hot stars emit most of their luminosity in blue or ultraviolet light; cool stars emit in the red or near-infrared. We can quantify the color of a star by comparing the flux in different wavelength bands – say, the blue band compared to the visible band, and this is known as a color index, for example

$$B - V = 2.5 \log(F_V / F_B) \qquad (A1.4)$$

is called the B–V color index. A larger color index means that the object is redder; a smaller color index means that it is bluer; so a relatively hot star has a smaller B–V color index than does a cool star. The color index is an intrinsic property of an object – independent of distance – unlike the apparent magnitude.

The luminosity of an object is, of course, also an intrinsic property, and is usually expressed by astronomers as an "absolute" magnitude. This is the apparent magnitude the object would have if it were placed at a standard distance, taken to be 10 parsecs or about 3×10^{17} m (more on parsecs below). By extragalactic standards this is a very small distance, so the absolute magnitudes of galaxies turn out to be quite large *negative* numbers: $M_G \approx -18$ to -23. The luminosity of an object compared to the solar luminosity is given by

$$M_G - M_\odot = -2.5 \log(L_G / L_\odot) \qquad (A1.5)$$

where $M_\odot$ is the absolute magnitude of the Sun (about -5.5 in the B band) and $L_\odot$ is the luminosity of the Sun (4×10^{33} ergs/s). The luminosities of galaxies typically range between 10^8 and $10^{11} L_\odot$.

I will frequently use the term "surface brightness". The surface brightness of an object is the flux emitted by the object directly at the source. For example, the Sun emits about 7×10^{20} ergs every second from every square centimeter of its surface. This would be the surface brightness of the Sun, and it is also an intrinsic property; it does not vary with distance. This may seem like a large value, but the solar surface brightness is actually only about 1% of a modern laser.

Stars not only emit continuum electromagnetic radiation; we can also detect radiation at distinct wavelengths, i.e., spectral lines. Sometimes these lines are seen in emission – enhanced radiation over the continuum – and sometimes in absorption – a deficiency of radiation at some wavelength. These lines are emitted by different atoms or molecules at definite wavelengths; for example, the H_α line of neutral hydrogen always appears at 6463 angstroms in the laboratory. However, if the line is arising in an object that is moving toward or away from us, this wavelength will be shifted by the well-known Doppler effect. The change in wavelength is proportional to the speed toward or away from the observer:

$$\Delta\lambda/\lambda = v/c \qquad (A1.6)$$

where c is the speed of light and v is the velocity toward or away from us (this formula is valid only when v is much less than c). If the object is moving toward us, the spectral line is shifted to the blue; away from us, it is shifted to the red. The Doppler effect is an extremely important tool in astronomy; it allows us to measure the velocity of an astronomical object toward or away from us as well as the rotation velocity in an extended object such as a galaxy.

A2 Distance in astronomy

If we look up on a clear night far away from a large city, we see that the sky is filled with stars. These stars all belong to the Milky Way Galaxy, a vast flattened disk-like system of visible stars and gas rotating about its center. The plane of this disk is apparent as the luminous band running across the sky, particularly noticeable in northern summertime. The distances to the stars that we can see are measured in units called "parsecs". As the Earth moves around the Sun, our perspective in viewing the sky changes and this causes an apparent change in the position of the stars (this can only be noticed for very nearby stars). The distance of a star whose position changes by one arc second (1/3600 of a degree) is defined as one parsec (there are actually no stars this close by) and it corresponds to 3.086×10^{13} kilometers (or about 3 light years, the distance light travels in one year). The stars we see with our unaided eye are mostly within distances of 200 parsecs (abbreviated pc). Galaxies have sizes of thousands of parsecs; thus it is useful to introduce the unit of kiloparsec (kpc) when discussing distances within galaxies. The Sun, for example, is about 8 kpc (8000 pc) from the center of the Milky Way. The nearest large galaxy, the great spiral galaxy in Andromeda, M31, is at a distance of about 700 kpc; most galaxies are much further. So, for extragalactic astronomy it is preferable to describe distances in a unit of a million parsecs, a megaparsec or Mpc. Relatively nearby galaxies are at distances of up to 20 Mpc; the most distant galaxies we observe are at distances of hundreds of Mpc.

One of the most significant astronomical discoveries of human history was the realization, in 1929, by the American astronomer Edwin Hubble (along with Milton Humason), that the light from distant galaxies was generally redshifted, that is, the galaxies are rushing away from each other with a speed that is proportional to the distance separating them. This has become immortalized as the Hubble law:

$$v = H_0 D \qquad (A2.1)$$

where v is the velocity of recession, D is the distance, and H_0 is the constant of proportionality – the "Hubble constant". If we can measure the velocity of galaxies via the Doppler shift and determine the distance by other means, for example, by looking at the apparent brightness of stars with known absolute luminosity, then we can calibrate this relation; we can measure the Hubble constant, which is obviously a number of profound cosmological significance. It has taken many decades, considerable controversy and, finally, the launch of the Hubble space telescope to arrive at the present value of the Hubble constant which is about 72 km/s/Mpc. That is to say, for every megaparsec in distance, the recession velocity of a galaxy is larger by 72 km/s.

Knowing the Hubble constant and measuring the recession velocity of a galaxy from its redshift, we can turn this relationship around and determine its distance. This is the principal means of measuring distance in extragalactic astronomy. For example, a galaxy that is receding with a velocity of 1000 km/s would be, from eq. A2.1, at a distance of 14 Mpc.

It is also evident that the inverse of the Hubble constant, $1/H_0$, has units of time – the Hubble time. In fact it is, naively, the time at which all galaxies were overlapping, or the age of the Universe. This turns out to be about 14 billion years. We can also determine the distance at which the recession velocity is equal to the speed of light and that is c/H_0 or 4000 megaparsecs – the Hubble distance. Beyond this radius we can no longer see galaxies because they are receding too fast for their light to reach us, so this describes a horizon for the Universe. Because the speed of light defines a causally connected region, the horizon is the scale over which events can influence each other.

A3 Galaxies

There are billions of galaxies out to a Hubble radius, but they are not distributed smoothly through space; they are clumped together in groups and clusters of galaxies. There are even clusters of clusters, superclusters, which have the morphologies of large walls or filaments of galaxies. Clusters of galaxies appear to be the largest gravitationally bound objects, and they are supported against their self-gravity by the random motion of the individual galaxies. In fact, we now know that clusters also contain hot X-ray emitting gas and that the total mass of this gas typically exceeds the mass of stars in galaxies. So in a sense, rich clusters of galaxies appear to be spheres of hot gas, rather like stars, supported against gravity by pressure forces.

Galaxies come in two basic flavors: there are the large disk galaxies like our own – flattened systems which are rotating about their centers – so-called spiral galaxies; and rounder or spheroidal galaxies which evidence little rotation – elliptical galaxies (see Figs. A3.1 and A3.2). The Milky Way Galaxy

Fig. A3.1. A spiral galaxy, actually a disk, seen here face-on. Visible are the bright regions of ionized gas surrounding bright young stars (recently formed out of the gas) as well as dark dust lanes arranged in the conspicuous spiral pattern. Because of the recent star formation and the hot young stars, these systems appear to be quite blue. The mass-to-light ratio of the stellar population is of the order of one in solar units.

Fig. A3.2. An elliptical galaxy appears rather dull compared to a spiral. These galaxies are flattened but not like a disk. The support against gravity is provided primarily by random motion of the stars rather than rotation. There is little gas and no evidence for recent star formation. The color of these systems is more red than that of spirals because of the old population of low-mass, cooler stars and no recent star formation. The mass-to-light ratio of the stellar population is higher than that of spirals, perhaps three to five in solar units.

contains about 100 billion (10^{11}) stars; its mass in stars would be a few times 10^{10} solar masses. The mass of the Sun is a standard unit of mass (2×10^{30} kilograms) in astronomy. The mass of the Galaxy, therefore, would be of the order of 10^{41} kilograms. Galaxies in general range between 10^7 to 10^{12} solar masses (abbreviated $M_\odot$), at least in visible stars.

Spiral galaxies contain gas as well as stars, although the mass of gas is typically only a few percent of that of stars – typically, but not always. There are some smaller galaxies where the mass of gas can actually dominate the observable mass of the system. This gas consists largely of neutral hydrogen (about 76%) along with 24% helium produced when the Universe was a few minutes old. In this interstellar medium there are also trace amounts of elements heavier than helium as well as solid particles (dust) consisting primarily of carbon, silicon and water ice and having sizes of the order of the wavelength of light. In the Milky Way this dust obscures the light from stars further away than a few tens of parsecs, but radio emissions from the gas come through to us undimmed. In some galaxies much of the hydrogen may be in molecular form – two hydrogens bound together to form the H_2 molecule.

Because of the gas, stars are still actively forming in spiral galaxies. A number of these stars are massive, hot and luminous and emitting radiation primarily in blue light (a relatively low B–V color index). Elliptical galaxies, on the other hand, contain very little gas, and star formation is not actively proceeding. These objects therefore consist primarily of old stars which are low-mass and emitting much of their light in the red part of the spectrum (a relatively high B–V spectral index). While the disks of spiral galaxies appear to be held up against their self-gravity primarily by rotation – the "centrifugal force" – elliptical galaxies are supported largely by the random motion of the stars – "pressure support".

An important property of an astronomical system such as a galaxy, especially in the context of dark matter, is the mass-to-light ratio, or M/L. For the Sun, this would be 0.5 ergs per second per gram. Typically, the mass-to-light ratio of a stellar system or a population of stars is expressed in solar units; e.g., if we say the mass-to-light ratio of a galaxy is five (M/L = 5), this means that the mass-to-light ratio is five times larger than the Sun – five times more mass per unit luminosity than for the Sun. Massive stars (say 10 times more massive than the Sun) have a very high luminosity – 1000 to 10 000 times more than the Sun, so the mass-to-light ratio for such objects is less than 0.001. However, stars with mass less than the Sun have much lower luminosity and high mass-to-light ratios (>1). Therefore, the mass-to-light ratios of the stellar component (the "stellar population") in elliptical galaxies is generally higher (3–5) than it is in spiral galaxies (1–2). A mass-to-light ratio in excess of 10 or so is difficult to achieve with normal stellar populations and would suggest the presence of unseen (non-stellar) matter.

A typical bright galaxy, spiral or elliptical, has a surface brightness of several hundred solar luminosities per square parsec. That means that every square parsec of a galaxy is emitting perhaps $100\,L_\odot$. In the past 20 years, a large population of low-surface-brightness galaxies has been discovered. These may have a surface brightness that is 10 times smaller ($10\,L_\odot/\mathrm{pc}^2$). If we know the mass-to-light ratio of a population of stars in a galaxy, we may also estimate the surface density of stellar mass in the galaxy. For example, if $M/L = 3$, the surface density might be $300\,M_\odot/\mathrm{pc}^2$.

A4 Weighing galaxies and clusters

I have already used the word "gravity" several times. Everyone has an intuitive sense of gravity – it is the force of attraction between objects with mass. In fact, it is one of the four basic forces of nature, the other three being the electromagnetic force (the attraction of metal objects for a magnet or of positive charges for negative charges); and the "weak" force and "strong" force which act only on subatomic distances (10^{-14} cm or less). These nuclear forces, being very short range, play no role in determining the dynamics of astronomical systems. The electromagnetic force is a long-range force but acts primarily between particles or objects with positive or negative charge. On a large scale the Universe is electrically neutral so the electromagnetic force also plays no role in celestial mechanics. The only remaining long-range force which can influence the motion of astronomical objects is the force of gravity.

The modern theory of gravity is general relativity written down by Albert Einstein in 1915. General relativity was preceded by special relativity (by 10 years) which posits that the speed of light is the same in all reference frames in constant relative motion with respect to one another. A well-known consequence of this proposal is the equivalence of matter and energy encapsulated by the famous formula

$$E = mc^2. \tag{A4.1}$$

Because of this equivalence of mass and energy, we know that electromagnetic radiation gravitates. Moreover, it has led to the convention of expressing the masses of subatomic particles in units of energy: for example, the mass of an electron is equivalent to an energy of 0.5 MeV (million electron volts) and that of a proton to roughly 1000 MeV (or 1 GeV). By Einstein's formula this would correspond to 1.7×10^{-24} grams. In Chapters 6 and 11 I will discuss hypothetical dark matter particles with masses of, perhaps, 100 GeV (100 times the proton mass).

General relativity addresses the motion of relativistic particles (such as photons) in a gravitational field. It relates the gravitational field to the curvature of four-dimensional space–time, and describes how that curvature is related to the distribution of mass–energy. In the limit of weak gravitational fields, as in the Solar System or the Galaxy, general relativity reduces to Newtonian gravity, which we consider further below. Newtonian gravity is thought to be sufficient in most astronomical contexts (apart from cosmology or in the vicinity of extremely compact objects such as neutron stars or black holes).

For an object like a star or a galaxy, the gravity force acts to pull the matter into the center. If the force were not resisted, then all objects would collapse to black holes which, of course, is not the case. For a star, the gas is hot and exerts an outward pressure, so it is the pressure force that resists the inward pull of gravity. The same is true for an elliptical galaxy, but now it is a gas of stars that exerts the pressure. And the same is true of a cluster of galaxies where the "gas" is the actual hot X-ray emitting gas as well as a gas of galaxies.

The force of gravity between two objects of mass m_1 and m_2 and separated by distance R was first described by Isaac Newton, and is given by

$$F = Gm_1m_2/R^2 \qquad \text{(A4.2)}$$

where G is the universal constant of gravity. Einstein's theory of gravity, general relativity, is the more complete theory, but on the scale of the Solar System and larger adds only very tiny corrections to Newton's original formula.

How does an object react when a force such as gravity is applied? This is encapsulated by Newton's famous second law of motion:

$$F = ma \qquad \text{(A4.3)}$$

where a is acceleration. F and a are vectors; they possess the attribute of direction as well as magnitude. So when a force is applied, the object accelerates; it changes its speed and/or direction. Eqs. A4.2 and A4.3 may then be applied to determine the acceleration due to gravity (the force per unit mass) on object m_1 at distance R from object m_2:

$$a = Gm_2/R^2. \qquad \text{(A4.4)}$$

Therefore, if we can measure the gravitational acceleration at distance R in such an object, we can determine its mass. For a particle, like a planet, in circular orbit around a central mass, like the Sun, the acceleration is the well-known "centripetal acceleration"; the acceleration that is present because the planet is always changing its direction (if there were no force the motion would be along a straight-line path). This centripetal acceleration is given by V^2/R where V is the velocity of circular motion. If we say that this centripetal acceleration is due to the force of gravity (and it must be for astronomical objects) then we have

$$V^2/R = GM/R^2 \tag{A4.5}$$

(where I have replaced m_2 by M). So if we measure V and R then we can determine the mass of the central object, and that would be

$$M = V^2 R/G. \tag{A4.6}$$

This, in effect, is how the mass of the Sun was originally measured – from the motion of planets about the Sun. If we are discussing the circular motion of particles about a central object we have

$$V = \sqrt{GM/R}. \tag{A4.7}$$

That is to say, the rotational velocity falls off as $1/\sqrt{R}$. In other words, the rotation law of the planetary system of the Sun is $V \propto 1/\sqrt{R}$. Because this was essentially first described by Kepler, such a rotation law is called Keplerian (although Kepler described it in terms of a period of revolution rather than orbital velocity).

If the mass has a spherical shape (spherically symmetric) and is not all concentrated in the center but is extended we can still use eq. A4.6 above to determine the mass inside radius R, that is $M(R)$; we just substitute $M(R)$ for M in the equation. But now, the rotation law is no longer Keplerian but depends upon the exact form of $M(R)$, the mass distribution in the object. We expect this to generally be the case in an extended object such as a spiral galaxy where we can measure the rotation velocity $V(R)$ at some radius (via the Doppler shift). In fact, the equation should be modified to

$$V(R) = p\sqrt{GM(R)/R} \tag{A4.8}$$

where p is a factor (of the order of one) that takes into account that the mass distribution in a galaxy may not be spherically symmetric (it certainly does not appear to be in spiral galaxies). Properly speaking p should also be a function of R. So if we can measure the rotation law of a spiral galaxy, the "rotation curve", we can, in principle, determine the mass distribution.

We can also estimate the total mass of a system like a cluster of galaxies or an elliptical galaxy which does not exhibit pure rotation. In this case it is useful to apply a very powerful theorem in classical mechanics, the virial theorem. A gravitating system has two kinds of energy: the kinetic energy or the total energy of motion of all the objects comprising the system; and the gravitational potential energy which is the energy required to pull the object apart and disperse it to an infinite distance. The kinetic energy, T, is given by

$$T = \frac{1}{2}MV^2 \tag{A4.9}$$

where M is the total mass of the system and V is the average velocity of objects in the system. The potential energy, taken to be negative, is given by

$$U = -GM^2/R \tag{A4.10}$$

where R is a characteristic radius or size of the system. The virial theorem tells us that for a system in equilibrium – one that is not collapsing or flying apart – these two kinds of energy are comparable; to be more precise:

$$T = -U/2 \tag{A4.11}$$

or

$$M = V^2R/G. \tag{A4.12}$$

This looks identical to eq. A4.6, but here remember that the virial theorem is global: V is the velocity dispersion or velocity spread of the objects comprising the system (individual stars in the case of a galaxy or individual galaxies in the case of a cluster of galaxies), R is the characteristic radius of the system, and M is the total mass.

A5 Cosmology

Cosmology is the study of the structure and evolution of the Universe as a whole. We can write down equations for the evolution of the Universe only by making a very powerful and restrictive assumption known as the "cosmological principle". The content of the cosmological principle is this: the Universe appears isotropic (it looks the same in all directions) and homogeneous (properties such as density or temperature do not vary with position) to any observer. As an assumption, the cosmological principle cannot be proven. But no observation performed so far is inconsistent with this assumption; at least no observation of the distant Universe. The density of galaxies appears to be independent of direction; the intensity of the "cosmic background radiation" (the microwave radiation from the early hot Universe) is highly isotropic; there is no significant variation with direction.

To derive the equations of the evolution of the Universe, we should combine the cosmological principle with the complete theory of gravity – general relativity. But it turns out that we can get the same result just by considering the motion of an expanding uniform sphere of cold gas (no pressure) in the present Universe acted upon only by Newtonian gravity. The escape velocity at the edge of this sphere is just given by

$$V_e^2 = 2GM/R \qquad (A5.1)$$

where M is the mass of the sphere and R is its radius. As specified by the cosmological principle, the sphere must be uniform, which means that it has a constant density ρ: therefore,

$$M = \frac{4\pi}{3}\rho R^3. \qquad (A5.2)$$

Then eq. A5.1 becomes

$$V_e^2 = \frac{8\pi}{3}G\rho R^2. \qquad (A5.3)$$

Because of the cosmological principle, the sphere must be expanding uniformly, which means that its shape does not change from spherical and maintains a uniform density as it expands. This is only possible if it expands equally in all directions and according to a Hubble law

$$V = HR. \qquad (A5.4)$$

(Interesting that the cosmological principle requires the Hubble law, but it does not say anything about H which may be positive, negative or zero.) We know that if the expansion velocity exceeds the escape velocity $V > V_e$ the sphere will expand forever, but if $V < V_e$ the sphere will eventually re-collapse. Therefore, a critical value of the expansion velocity is just $V = HR = V_e$; this is the expansion velocity such that the sphere will just expand to infinity. Substituting this into eq. A5.3 we have,

$$H^2R^2 = \frac{8\pi}{3}G\rho R^2. \qquad (A5.5)$$

Conveniently, the R^2 cancels out of the equation and, with a little algebra, we can define a critical density relevant to the present Universe ($H = H_0$)

$$\rho_c = \frac{3H_0^2}{8\pi G}. \qquad (A5.6)$$

If the density of the sphere (the Universe) at the present time is smaller than this critical value, the Universe will expand forever; if it is larger, the Universe will re-collapse. With $H_0 = 72$ km/s/Mpc this critical density is about 10^{-29} g/cm^3 or about 10 hydrogen atoms per cubic meter. This is an average density: it would be the density of the Universe if all the matter were spread out uniformly.

Cosmologists usually write the average density in terms of the critical density, and that parameter is called Ω_0; that is,

$$\Omega_0 = \frac{\rho}{\rho_c} \tag{A5.7}$$

(we say Ω_0 to specify this is the present value of Ω; this quantity may, like the Hubble parameter, evolve and will have a different value Ω in the past). If Ω_0 is less than one, the Universe will expand forever; if Ω_0 is larger than one, it will re-collapse.

A simple cancellation of R^2 in equation A5.5 gives

$$H^2 = \frac{8\pi G}{3}\rho. \tag{A5.8}$$

This, relating the expansion rate to the density, is a simple version of the "Friedmann equation". In a cosmological context, the redshift of electromagnetic radiation, designated $Z = \Delta\lambda/\lambda$, may be viewed as the stretching of the wavelength due to the expansion of the Universe. Then, considering our spherical piece of the Universe, its radius was smaller at earlier times or higher redshift: $R \propto 1/(1 + z)$. In an expanding universe, the mass density of ordinary non-relativistic particles, ρ, decreases as the number density of particles and this varies with the inverse volume $\rho \propto 1/R^3 \propto (1 + z)^3$. So with a little bit of algebra and using the definition of Ω_0, we can write the Friedmann equation as

$$(H/H_0)^2 = \Omega_0(1 + z)^3. \tag{A5.9}$$

Obviously, since $H = H_0$ at present ($z = 0$) this formula is only strictly valid for $\Omega_0 = 1$. More generally we may write

$$(H/H_0)^2 = \Omega_m(1 + z)^3 + \Omega_k(1 + z)^2 \tag{A5.10}$$

where Ω_m is the present density parameter of ordinary matter and Ω_k describes the curvature of the Universe. This constant, Ω_k, may be positive (negative curvature), negative (positive curvature) or zero (flat universe). Again the boundary condition at $z = 0$ requires that $\Omega_m + \Omega_k = 1$.

Cosmologists have always taken $\Omega_k = 0$ to be a preferred value, and this is because of the nature of the Friedmann equation. If the early Universe were only very slightly negatively curved, it would now be essentially empty, $\Omega_m \approx 0$ (or $\Omega_k = 1$). If, on the other hand, the Universe were only slightly positively curved, it would have re-collapsed long before it reached its present age. But if $\Omega_k = 0$ quite precisely, then $\Omega_m = 1$ always. This has for decades been considered as a fine-tuning problem in cosmology: how is it that Ω_k is so close to zero? Now there is a physical theory that explains this fine tuning: "inflation". This theory is based upon concepts in particle physics and the unification of forces and implies that there is an epoch in the extremely early Universe (the exact age depends upon the mechanism driving inflation but 10^{-34} seconds is reasonable) when the Universe went through a period of exponential expansion which drove $\Omega_k \to 0$ to extremely high precision. One could say that the radius of the curved Universe increased so enormously that it appears flat (just as the Earth, large compared to us, appears flat). So, in a sense, $\Omega_k = 0$ is the natural and predicted value.

Inflation also smooths any wrinkles in the fabric of space–time; it drives the Universe to a highly isotropic and homogeneous state – one could say, it underlies the cosmological principle. But this process of exponential expansion also generates very small density fluctuations in the hot fluid – fluctuations which are the seeds of future structure.

The matter density of the Universe ρ may have several components. One of these is certainly visible matter, the matter in the form of stars in galaxies, Ω_v. When cosmologists carry out a census of all the shining matter in stars locally, they find

$$\Omega_v \approx 0.003.$$

If all the matter of the Universe were in the form of visible stars, then there would be insufficient gravity to cause the Universe to re-collapse. But I mentioned that there is also hot gas detected in clusters of galaxies. This hot gas could amount to

$$\Omega_g \approx 0.0025.$$

So altogether, the density in directly-detected mass is on the order of

$$\Omega \approx \Omega_v + \Omega_g \approx 0.005,$$

or far short of the preferred value of $\Omega_m = 1$.

A6 Radiation and the thermal history of the Universe

In the early 1960s, Arno Penzias and Robert Wilson working at the Bell Telephone Laboratory developed a sensitive microwave (radio) receiver attached to a horn-type antenna designed to detect signals only from the sky and not from the ground. They had a very good understanding of their receiver and antenna and when they eliminated all known sources of noise, they found a mysterious remaining background signal that seemed to be coming from all directions in the sky. This signal was equivalent to that which would be emitted by an object at the extremely low temperature of 3 kelvins – 3 degrees above absolute zero (Penzias and Wilson, 1965). As it turned out, this signal was from the Universe as a whole; it is the remnant radiation from an earlier stage in the evolution of the Universe: the hot Big Bang Universe. This was a discovery of fundamental importance, and it proved that the Universe was once much hotter and denser. It confirmed a cosmology proposed years before by George Gamow and his collaborators (Chapter 6).

Of course, electromagnetic radiation has energy and, therefore, also mass (remember $E = mc^2$). So the primordial radiation also makes a contribution to the present mass density of the Universe. This mass density is proportional directly to the fourth power of the temperature and, in terms of the critical density, turns out to be

$$\Omega_r = 5 \times 10^{-5}.$$

So, as far as the gravity of the Universe is concerned, electromagnetic radiation (photons) is quite unimportant at present (this was not always true, as we shall see below).

The cosmic microwave background photons detected by Penzias and Wilson come from all directions in the sky and have an energy distribution that is characteristic of a perfectly radiating and absorbing object (a black body) at a temperature of about 2.732 kelvins. But the radiation has not always been so cool. The Universe is expanding, which means that in the past it was smaller and denser and warmer. In order to understand why the concept of dark matter has also emerged in the context of what is now the standard Big Bang cosmology it is necessary to consider the thermal history of the early Universe as implied by the existence of the background radiation, and in particular, the formation of structure from an initially hot, extremely homogeneous expanding medium.

When astronomers look at distant galaxies, they are seeing these objects as they were, not as they are: they are looking into the past. This is because of the finite time that it takes for light to travel to us from an object which is many megaparsecs away (one megaparsec means 3 million years). So, the larger the redshift of an object, the further back in time we are looking. For example, at a redshift of two, the look-back time is 10 billion years, almost 75% of the age of the Universe.

Returning to our spherical piece of the Universe that is expanding along with the rest of the Universe, the temperature of radiation in that expanding sphere decreases as the radius of the sphere increases, $T \propto 1/R$. This means that in the past the radiation temperature, in terms of the present temperature, T_0, depends upon redshift as

$$\frac{T}{T_0} = 1 + z. \tag{A6.1}$$

So at a redshift of two, the radiation temperature was three times higher than it is at present.

We have seen that the mass density of ordinary non-relativistic matter, ρ_m, simply decreases as the inverse volume, or

$$\frac{\rho_m}{\rho_{m0}} = (1 + z)^3, \tag{A6.2}$$

where ρ_{m0} is the density of matter in the Universe at the present time. The number density of cosmological photons in the Universe also decreases with increasing redshift as $(1 + z)^3$, but eq. A6.1 is telling us that the energy (or equivalently the mass) of each photon varies as $(1 + z)^1$. Therefore, the equivalent mass density of radiation, ρ_r, increases with redshift as

$$\frac{\rho_r}{\rho_{r0}} = (1 + z)^4, \tag{A6.3}$$

where ρ_{r0} is the present mass density of photons (this would be true of any fluid consisting of highly relativistic particles, such as massless neutrinos).

Knowing the redshift dependence of the radiation mass density, we may now also add radiation to the Friedmann equation which becomes

$$(H/H_0)^2 = \Omega_r(1 + z)^4 + \Omega_m(1 + z)^3 + \Omega_k(1 + z)^2. \tag{A6.4}$$

The mass density of radiation increases faster with redshift than does that in ordinary matter, and this implies that at some point in the past, beyond a critical redshift z_c, the mass density of radiation dominated the mass density of non-relativistic matter. That critical redshift, given by the condition $\rho_m = \rho_r$, is

$$z_c = 22\,000\Omega_m, \tag{A6.5}$$

where Ω_m is the density parameter of non-relativistic matter (here I have taken $H_0 = 72$ km/s/Mpc, its current preferred value). In other words if $\Omega_m \approx 1$ then radiation would dominate over matter at redshifts higher than 22 000. This would correspond to an age for the Universe of about 1000 years and, from eq. A5.1, a temperature of 60 000 kelvins. At higher temperatures, or earlier times, the Universe could be described as a radiation universe – matter is gravitationally unimportant.

At such a high temperature there could be no neutral atoms. Apart, possibly, from exotic dark matter, there would be free electrons, protons and helium nuclei (protons and neutrons are called baryons, so this would be baryonic matter). All of the baryonic matter in the Universe, mostly hydrogen and helium, would be completely ionized. But when the temperature of the Universe falls to approximately 3000 degrees (at a redshift of about 1000), the protons and electrons have sufficiently low kinetic energy that they can combine and form neutral hydrogen. Before this event, the radiation and baryonic matter are tightly coupled – the photons cannot move freely (free-stream) but are continually scattered by the free electrons. It is, in a sense, a single fluid – a baryon–photon fluid. The joining of protons and electrons to form neutral hydrogen at $z = 1000$ releases the photons and they can freely stream at the speed of light. All of the CMB photons reaching us now are coming from this opaque wall at a redshift of 1000. This epoch of hydrogen-atom formation is called "decoupling" because that is when the photons decouple from the baryons and become, in effect, a separate fluid.

An aspect of the hot Big Bang model originally proposed by Gamow and his student Alpher, is that light elements such as helium and lithium were produced by fusion of hydrogen when the Universe was very young – a couple of minutes – and very hot – about one billion degrees (Gamow and Alpher actually proposed that all elements heavier than hydrogen were produced in the Big Bang; an idea that we now know is incorrect). The hydrogen nucleus is simply one proton. The helium nucleus consists of two protons and two neutrons so, basically, two protons had to fuse with two free neutrons to produce helium. But the free neutrons (not found in an atomic nucleus) are unstable and decay within 10 minutes, so this had to happen before the Universe was 10 minutes old and all the free neutrons disappeared.

Most of the mass of atoms is in the form of protons and neutrons (baryons), which comprise the nuclei of all atoms (electrons are 2000 times lighter than baryons). Therefore most of the ordinary detectable matter of the Universe – stars, interstellar gas, planets, humans – is in the form of baryonic matter. The total density of baryonic matter determines how much helium and other light elements, such as deuterium (an isotope of hydrogen consisting of one proton and one neutron), lithium and beryllium, are produced in the early Universe. So, in principle, if astronomers can measure the original (primordial) abundance of these elements, then they can determine the total density of baryons in the Universe. Measuring abundances is not easy or straightforward; these have to be primordial, that is, the abundance before matter was further processed in stars, and this problem remains an active area of research. The results point to a primordial abundance of helium of about 24% and a much smaller abundance of deuterium. But these observations are sufficient to constrain the baryonic density of the Universe, and it is

$$\Omega_b = 0.04 \text{ to } 0.05.$$

This is a very important result with enormous consequences for the concept of dark matter and its required properties, as we shall see.

These events in the early history of the Universe are illustrated schematically in Fig. A7.1 which shows a time sequence with the temperature of the radiation field also indicated. Major events such as the nucleosynthesis of deuterium and helium, the dominance of matter over radiation (depending, of course, on Ω_m) and decoupling of photons from baryons are indicated.

Recall that the detectable baryonic matter in the Universe (in stars and hot gas) is $\Omega_v + \Omega_g \approx 0.005$. But $\Omega_b \approx 0.05$ then means that only one-tenth of the baryons are actually shining; i.e., observed in visible stars or emitting X-rays as hot gas in clusters. In other words, 90% of the baryons have not yet been detected; one could say that they are dark baryons. But even so, we see that the density of baryons is still only 5% of that required to bind the Universe; so, with baryons alone, the Universe is unbound and "open". But there are other possibilities.

A7 Non-baryonic matter and dark energy

Not all matter in the Universe consists of baryons. Electrons are not baryons; negatively charged electrons normally combine with positively charged nuclei to form neutral atoms, but electrons comprise a very small fraction of the mass – 1/2000. Neutrinos are another kind of almost massless particle emitted in certain kinds of nuclear reactions – β decays. Until fairly recently they were thought to be absolutely massless, but now we know they are not; they have a mass, expressed in units of energy, of at least 0.05 electron volts (recall that the mass of an electron by comparison is 500 000 eV and the mass of a proton is one billion eV).

There are three forms or flavors of neutrinos associated with other kinds of particles: electron neutrinos, muon neutrinos and tau neutrinos. We now know that the neutrinos change form: a muon neutrino can turn into an electron neutrino and back again. This is only possible if the neutrinos have mass. But these "oscillations" provide only a lower limit on the mass; at least one sort of neutrino must be more massive than 0.05 eV.

Even though the neutrinos have this extremely low-mass, there are copious quantities of them. Neutrinos, like photons, are also produced in the early hot Universe. In fact, there are about as many neutrinos as photons, which means that they are a non-negligible component of the mass density of the Universe: $\Omega_v \geq 0.003$.

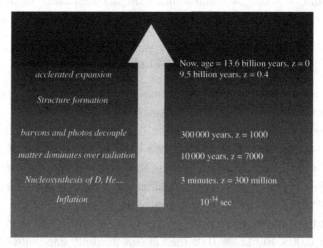

Fig. A7.1. A timeline showing major events in the history of the Universe.

So there is at least as much mass of neutrinos as of visible baryons, and there could be more. Neutrinos do not emit radiation so they could be called dark matter – non-baryonic dark matter. This is a term that is discussed in Chapter 6, but the point is that such dark matter certainly exists and is a component of the Universe.

The mass of the neutrino can, in principle, be measured in the laboratory. The nuclei of certain atoms, such as the tritium isotope of hydrogen (one proton, two neutrons), are unstable to β decay; i.e., one of the neutrons decays spontaneously to a proton, that remains in the nucleus, an electron and an electron neutrino. In tritium β decay the energy of this process is 18.6 keV which is shared between the electron and the neutrino. By carefully measuring the energy distribution of the electrons near this cut-off, one can, in principle, determine the mass of the neutrino electron (it carries away at least its rest energy). It is a difficult experiment, and, so far, there are only upper limits on this mass: $m_\nu < 2\,\text{eV}$. This means that the maximum contribution of neutrinos to the density of the Universe would be $\Omega_\nu \approx 0.1$, again hardly enough to close the Universe (with baryons the total matter density would amount to $\Omega = 0.15$).

There is another possibility; a very strange form of matter not connected with a particle at all, but with energy – so-called "dark energy". When Einstein wrote down his famous field equations of general relativity, the Universe was thought to be static – not expanding or contracting. Einstein realized that he could not achieve a static state with matter alone – the Universe had to be either expanding or contracting – so he added an additional term to his equations: the so-called cosmological constant. The effect of the cosmological constant is to add a universal repulsion in the Universe that increases directly with distance. This repulsion could balance the normally attractive force of gravity and make a static universe possible. (Later when it was found by Hubble that the Universe really was expanding, Einstein referred to the cosmological constant as his greatest blunder.)

But now the cosmological constant has re-emerged in a somewhat different guise, as "vacuum energy density". Modern physics theory, quantum field theory, tells us that the vacuum is not empty, but seething with virtual particles which pop into and out of existence. This virtual fluid has a pressure which is actually negative, and the effect of this negative pressure is to promote a universal repulsion, just like Einstein's cosmological constant. So the cosmological constant is now often viewed as a representation of this vacuum energy density, although, in the particle-physics context the vacuum energy density is far too large to be identified with the cosmological term in Einstein's equation (this is a fundamental problem of modern physics). In the present jargon this is called "dark energy", and it does not dilute in the same way as matter does as the Universe expands; it does not dilute at all if it is really like the cosmological constant. We should now add this non-diluting substance to the Friedmann equation:

$$(H/H_0)^2 = \Omega_r(1+z)^4 + \Omega_m(1+z)^3 + \Omega_k(1+z)^2 + \Omega_\Lambda \qquad (A7.1)$$

where the final term represents the cosmological term. This promotes an accelerated expansion of the Universe, and a very new astronomical result is that the Universe really does seem to be undergoing accelerated expansion (Chapter 9). So apparently some fraction of the mass density of the Universe actually is in the form of this vacuum energy density – this is now thought to be about 70%. In any case, from the form of this final equation, it is obvious that at earliest times (highest redshift), radiation will dominate the dynamics of the Universe, but later (lower redshift) matter will dominate and that this matter component will include a non-baryonic part as well as the usual baryons which compose everything that we see. Later still, curvature (if it is not zero) will dominate, and, ultimately, the cosmological term (if it is not zero) will dominate, and the Universe will expand exponentially. Again, the condition at $z = 0$ requires that $\Omega_r + \Omega_m + \Omega_k + \Omega_\Lambda = 1$.

A8 Gravitational instability and the growth of structure

In our picture of the early Universe, matter and radiation are initially distributed very smoothly. As discussed earlier, this is an aspect of the inflationary paradigm: this period of rapid exponential expansion in the very early Universe produces an extremely smooth, isotropic and homogeneous Universe, apart from the tiny density fluctuations on all size scales that are in fact quantum fluctuations generated in this explosively expanding medium. But in the present Universe, the matter distribution is far from homogeneous. We observe considerable structure on a range of scales: stars, galaxies, clusters of galaxies, a network of superclusters. How does such structure form?

Since the time of Newton it has been realized that an almost homogeneous medium is gravitationally unstable: a small density enhancement (a fluctuation) in an otherwise homogeneous medium will collapse due to its self-gravity. This was first considered in detail by Sir James Jeans in 1902 who derived a criterion for gravitational collapse. If the thermal pressure of a fluctuation is insufficient to counteract gravity, then the system will collapse. The collapse proceeds exponentially on a timescale of

$$t_c \approx \frac{1}{\sqrt{G\bar{\rho}}} \qquad (A8.1)$$

where $\bar{\rho}$ is the average density of the medium and G is the Newtonian gravitational constant. This timescale is also about equal to the age of the Universe when the average density is $\bar{\rho}$. If the density variation represented by a fluctuation is given by $\delta\rho$ (i.e., $\delta\rho = \rho - \bar{\rho}$) then the density fluctuation would grow as

$$\frac{\delta\rho}{\bar{\rho}} = e^{t/t_c}. \tag{A8.2}$$

If sound waves moving at a velocity c_s can propagate across some region with a size of l on a timescale less than the collapse scale t_c, then pressure forces can prevent collapse on that scale or smaller. Therefore, a critical length scale would be given by

$$l_c \approx \frac{c_s}{\sqrt{G\bar{\rho}}}. \tag{A8.3}$$

This is called the "Jeans length". For larger regions, the force of gravity overcomes the pressure forces resisting collapse. Only fluctuations larger than the Jeans length can collapse. Smaller fluctuations propagate as sound waves.

When the baryonic matter is still ionized (at $T > 3000\,\mathrm{K}$) then the effective sound speed in the baryon–photon fluid is about c, the speed of light (actually $c_s = c/\sqrt{3}$). Because the collapse timescale is essentially the age of the Universe, this means that l_c would be the size of the horizon, a causally connected region, at that epoch. Therefore, this Jeans length is comparable to a causally connected region, and by the Jeans criterion only regions larger than a horizon could collapse. But collapse is only possible over a causally connected region, which implies that no gravitational collapse is possible before decoupling. (This is not quite accurate: a positive density fluctuation larger than the horizon comprises, in a sense, a separate universe that expands more slowly than the Universe on average – the amplitude of the fluctuation $\delta\rho/\rho$ will slowly become larger.)

At decoupling ($z = 1000$) the sound speed in the baryon fluid drops dramatically, from about $170\,000\,\mathrm{km/s}$ to $5\,\mathrm{km/s}$, and gravitational collapse in this baryon component can begin over a wide range of scales smaller than the horizon. If $\Omega_m = 1$ in cold matter at the present epoch, then the average density of the Universe is $10^{-29}\,\mathrm{g/cm^3}$. At the epoch of decoupling, the density would be larger by a factor of 10^9 (eq. A6.2). Then from eq. A8.1 we find that, at that epoch, the timescale for collapse of structures on all scales larger than about $5\,\mathrm{pc}$ (the Jeans length) is about one million years; far less than the age of the Universe at 14 billion years. One would think that there should be no problem forming the presently observed structure in the Universe if collapse begins at decoupling (when the Universe is only about $300\,000$ years old). But, in fact, this is not the case. Jeans' original work on gravitational collapse applies to a static medium. But the universal fluid is not static; it is uniformly expanding and this changes the nature of the collapse.

The problem of gravitational instability in an expanding medium was first considered by the Russian physicist, Evgenii Lifshitz, in 1946. He realized that in such a case, the collapse is no longer exponential in time, but a power law; the collapse proceeds much less rapidly. Assuming that the present amplitude of a fluctuation is unity, i.e., $(\delta\rho/\rho)_0 = 1$, then the amplitude at higher redshifts is given by

$$\frac{\delta\rho}{\bar{\rho}} = 1/(1+z) \qquad\qquad (A8.4)$$

while the Universe is matter dominated.

At the present epoch, it must be the case that the density fluctuations, on the scale of galaxies or clusters, are comparable to or larger than the average density; after all, we see very condensed structure such as galaxies. This means that now $\delta\rho/\bar{\rho} \approx 1$ at least. Then, from eq. A8.4, the amplitude of the fluctuations at decoupling ($z = 1000$) had to be $\delta\rho/\bar{\rho} \approx 0.001$ in order to grow to the structures we see at present. Moreover, these fluctuations should be apparent as comparable variations in the temperature of the cosmic microwave background radiation across the sky, because before decoupling the baryons and photons were a single fluid. We should observe fluctuations in the temperature of the CMB at the level of about 10^{-4} to 10^{-3}, at least on an angular scale corresponding to galaxies or clusters of galaxies (1 arc minute to one-half degree). However, in the 25 years following the discovery of the CMB the upper limits on the temperature fluctuations were pushed down to below 10^{-4}, and nothing was seen. How is it then that the structure could possibly form? I consider this question in a historical context in Chapter 6.

References

Ahmed, Z. *et al.* (2009). Search for weakly interacting massive particles with the first five-tower data from the cryogenic dark matter search at the Soudan Underground Laboratory, *Phys. Rev. Lett.* **102**, 011301.

Ahronian, F. A., Hofmann, W., Konopelko, A. K. and Voelk, H. J. (1997). The potential of ground based arrays of imaging atmospheric Cherenkov telescopes, *Astropart. Phys.* **6**, 343–368.

Allen, R. J. and Shu, F. H. (1979). The extrapolated central surface brightness of galaxies, *Astrophys. J.* **227**, 67–72.

Alpher, R. A., Bethe, H. A., and Gamow, G. (1948). The origin of chemical elements, *Phys. Rev.* **73**, 803–4.

Alpher, R. A. and Herman, R. C. (1949). Remarks on the evolution of the expanding Universe, *Phys. Rev.* **75**, 1089–95.

Anderson, J. D., Laing, P. A., Lau, E. L., Liu, A. S., Nieto, M. M., and Turyshev, S. G. (1998). Indication, from Pioneer 10/11, Galileo, and Ulysses data, of an apparent anomalous, weak, long-range acceleration, *Phys. Rev. Lett.* **81**, 2858–61.

Aprile, E. *et al.* 2012. Dark matter results from 225 live days of XENON100 data, *Phys. Rev. Lett.* **109**, 181301.

Athanassoula, E. and Sellwood, J. A. (1986). Bi-symmetric instabilities of the Kuz'min/Toomre disc, *Mon. Not. Roy. Astron. Soc.* **221**, 213–32.

Babcock, H. (1939). The rotation of the Andromeda nebula, *Lick Obs. Bull.,* no. 498, Berkeley, Univ. of Calif. Press, pp. 41–51.

Bahcall, J. N. and Davis, R. (1976), Solar neutrinos, a scientific puzzle, *Science* **191**, 264–267.

Baugh, C. (2006). A primer on hierarchical galaxy formation: the semi-analytical approach, *Rep. Prog. Phys.* **69**, 3101–56.

Begeman, K. G. (1987). HI rotation curves of spiral galaxies, PhD dissertation, Univ. of Groningen.

Begeman, K. G. (1989). HI rotation curves of spiral galaxies, *Astron. Astrophys.* **223**, 47–60.

Behnke, I. E. *et al.* (2008) Spin-dependent WIMP limits from a bubble chamber, *Science* **319**, 933–6.

Bekenstein, J. D. (2004). Relativistic gravitation theory for the modified Newtonian dynamics paradigm, *Phys. Rev. D* **70**, 083509.

Bekenstein, J. D. and Milgrom, M. (1984). Does the missing mass problem signal the breakdown of Newtonian gravity?, *Astrophys. J.* **286**, 7–14.

Bekenstein, J. D. and Sanders, R. H. (1994) Gravitational lenses and unconventional gravity theories, *Astrophys. J.* **429**, 480–90.

Bernabei, R. *et al.* (2008). First results from DAMA/LIBRA and the combined results with DAMA/NaI, *Eur. Phys. J.* **C56**, 333–55.

Blumenthal, G. R., Faber, S. M., Primack, J. R., and Rees, M. J. (1984). Formation of galaxies and large-scale structure with cold dark matter, *Nature* **311**, 517–25.

Blumenthal, G. R., Pagels, H., and Primack, J. R. (1982). Galaxy formation by dissipationless particles heavier than neutrinos, *Nature* **299**, 37–8.

Bond, J. R., Efstathiou, G., and Silk, J. (1980). Massive neutrinos and the large scale structure of the Universe, *Phys. Rev. Lett.* **45**, 1980–4.

Bond, J. R. and Efstathiou, G. (1984). Cosmic background radiation anisotropies in universes dominated by non-baryonic dark matter, *Astrophys. J.* **285**, L45–L48.

Bond, J. R. and Szalay, A. S. (1983). The collisionless damping of density fluctuations in an expanding universe, *Astrophys. J.* **274**, 443–68.

Bondi, H. and Gold, T. (1948). The steady state theory of the expanding Universe, *Mon. Not. Roy. Astron. Soc.* **108**, 252–70.

Bosma, A. (1978). The distribution and kinematics of neutral hydrogen in spiral galaxies of various morphological types, PhD dissertation, The University of Groningen.

Bosma, A. (1981). 21-cm line studies of spiral galaxies. II. The distribution and kinematics of neutral hydrogen in spiral galaxies of various morphological types, *Astron. J.* **86**, 1825–46.

Broeils, A. H. (1992) Dark and visible matter in spiral galaxies, PhD dissertation, The University of Groningen.

Burbidge, M. E., Burbidge, G. B., Fowler, W. A., and Hoyle, F. (1957). Synthesis of elements in stars, *Rev. Mod. Phys.* **29**, 547–650.

Casertano, S. (1983). Rotation curve of the edge-on spiral galaxy NGC 5907: disk and halo masses, *Mon. Not. RAS* **203**, 735–7.

Casertano, S. and van Gorkom, J. (1991). Declining rotation curves – the end of a conspiracy?, *Astron. J.* **101**, 1231–41.

Chandrasekhar, S. (1941). The time of relaxation of stellar systems, *Astrophys. J.* **93**, 285–304.

Chang, J. *et al.* (2008). An excess of cosmic ray electrons at energies of 300–800 GeV, *Nature* **456**, 362–5.

Clowe, D., Bradač, M., Gonzalez, A. H., Markevitch, M., Randall, S. W., Jones, C., and Zaritsky, D. (2006). A direct empirical proof of the existence of dark matter, *Astrophys. J.* **648**, L109–L113.

Cowsik, R. and McClelland, J. (1973). Gravity of neutrinos of nonzero mass in astrophysics, *Astrophys. J.* **180**, 7–10.

de Bernardis, P. *et al.* (2000). First Results from the BOOMERanG Experiment, *Am. Inst. Phys. Conf. Proc.* **555**, 85–94.

de Lapparent, V., Geller, M. J., and Huchra, J. P. (1986). A slice of the Universe, *Astrophys. J.* **302**, L1–L5.

Dicke, R. H., Peebles, P. J. E., Roll, P. G., and Wilkinson, D. T. (1965). The cosmic black body radiation, *Astrophys. J.* **142**, 414–9.

Disney, M. J. (1976). Visibility of galaxies, *Nature* **263**, 573–5.

Duffy, L. D. *et al.* (2006). High resolution search for dark matter axions, *Phys. Rev. D* **74**, 012006.

Efstathiou, G. and Bond, J. R. (1986). Microwave background fluctuations and dark matter, *Phil. Trans. Roy. Soc. London, Series A, Math. Phys. Sci.* **320**, 585–94.

Emden, R. (1907), Gaskugeln, Teubner (Leipzig, Berlin).

Ewen, H. and Purcell, E. (1951). Observations of a line in the galactic radio spectrum; radiation from galactic hydrogen at 1,420 Mc/s, *Nature* **168**, 356.

Faber, S. M. and Gallagher, J. (1979). Masses and mass-to-light ratios of galaxies, *Ann. Rev. Astron. Astrophys.* **17**, 135–87.

Faber, S. M. and Jackson, R. E. (1976). Velocity dispersions and mass-to-light ratios for elliptical galaxies, *Astrophys. J.* **204**, 668–83.

Finzi, A. (1963). On the validity of Newton's law at a long distance, *Mon. Not. RAS* **127**, 21–30.

Freeman, K. C. (1970). On the disks of spiral and S0 Galaxies, *Astrophys. J.* **160**, 811–30.

Freeman, K. C. and McNamara, G. (2006). *In Search of Dark Matter*, Springer-Praxis (Berlin).

Friedmann, A. (1922). Uber die Kruemming des Raumes, *Z. Phys.* **10**, 377–38.

Gaitskell, R. J. (2004). Direct detection of dark matter, *Ann. Rev. Nuc. Part. Sci.* **54**, 315–59.

Garnavich, P. M. *et al.* (1998). Constraints on cosmological models from Hubble space telescope observations of high-z supernovae, *Astrophys. J.* **493**, L53–L57.

Gershtein, S. S. and Zeldovich, Ya. B. (1966). Rest mass of muonic neutrino and cosmology, *ZhETF Pis'ma* **4**, 174–5.

Gott, J. R., Gunn, J. E., Schramm, D. N., and Tinsley, B. M. (1974). An unbound Universe?, *Astrophys. J.* **194**, 543–53.

Gunn, J. E., Lee, B. W., Lerche, I., Schramm, D. N., and Steigman, G. (1978). Some astrophysical consequences of the existence of a heavy stable neutral lepton, *Astrophys. J.* **223**, 1015–31.

Hoekstra, H., Franx, M., Kuijken, K., and Squires, G. (1998). Weak lensing analysis of Cl 1358+62 using Hubble space telescope observations, *Astrophys. J.* **504**, 636–60.

Hohl, F. (1971). Numerical experiments with a disk of stars, *Astrophs. J.* **168**, 343–59.

Hohl, F. and Hockney, R.W. (1969). A computer model of disks of stars, *J. Comp. Phys.* **4**, 306–312.

Hoyle, F. (1948). A new model of the expanding Universe, *Mon. Not. Roy. Astron. Soc.* **108**, 372–82.

Hoyle, F. and Taylor, R. J. (1964). The mystery of the cosmic helium abundance, *Nature* **204**, 1108–10.

Hu, W. and Sugiyama, N. (1995). Toward understanding the CMB anisotropies and their implications, *Phys. Rev. D* **51**, 2559–630.

Jansky, K. G. (1933). Radio waves from outside the Solar System, *Nature* **132**, 66.

Jones, C. and Forman, W. (1984). The structure of clusters of galaxies observed with Einstein, *Astrophys. J.* **276**, 38–55.

Jungman, G., Kamionkowski, M., and Greist, K. (1996). Supersymmetric dark matter, *Phys. Rep.* **267**, 195–373.

Kahn, F. D. and Woltjer, L. (1959). Intergalactic matter and the galaxy, *Astrophys. J.* **130**, 705–17.

Kalnajs, A. J. (1983). *IAU Symp. 100: Internal Kinematics and Dynamics of Galaxies*, ed. E. Athanassoula, Reidel (Dordrecht), p. 87.

Kane, G. (2000). *Supersymmetry: Squarks, Photinos and Unveiling the Ultimate Laws of Nature*, Perseus Publishing (Cambridge, Mass).

Kent, S. M. (1986). Dark matter in spiral galaxies. I – Galaxies with optical rotation curves, *Astron. J.* **91**, 1301–27.

Klypin, A., Gottloeber, S., Kravtsov, A. V., and Khokhlov, A. M. (1999). Galaxies in N-body simulations: overcoming the overmerging problem, *Astrophys. J.* **516**, 530–51.

Klypin, A. A. and Shandarin, S. F. (1983). Three-dimensional formation of large scale structure in the Universe, *Mon. Not. RAS* **204**, 891–907.

Kuhn, T. S. (1962). *The Structure of Scientific Revolutions*, Univ. of Chicago Press (Chicago).

Lamaitre, G. (1927). Un Univers homogene' et de rayon croissant rendant des nebuleuses extra-galactique, *Ann. Soc. Sci. de Bruxelles* **A47**, 49–59.

Lifshitz, E. M. (1946). On the gravitational instability of the expanding Universe, *Journ. Phys. USSR* **10**, 116–22.

Lin, C. C. and Shu, F. H. (1964). On the spiral structure of disk galaxies, *Astrophys. J.* **140**, 646–655.

Lynds, R. and Petrosian, V. (1986). Giant luminous arcs in galaxy clusters, *Bull. Am. Astron. Soc.* **18**, 1014.

Mayall, N. (1951). Comparison of rotational motions observed in spirals M 31 and M 33 and in the Galaxy, *Pub. Obs. Michigan* **10**, 19.

McGaugh, S. S. and de Blok, W. J. G. (1998). Testing the hypothesis of modified dynamics with low surface brightness galaxies and other evidence, *Astrophys. J.* **499**, 66–81.

McGaugh, S. S., Schombert, J. M., Bothun, G. D. and de Blok, W. J. G. (2000). The baryonic Tully–Fisher relation, *Astrophys. J.* **533**, L99–L102.

Milgrom, M. (1983). A modification of Newtonian dynamics as a possible alternative to the hidden matter hypothesis, *Astrophys. J* **270**, 365–70.

Milgrom, M. (1984). Isothermal spheres in the modified dynamics, *Astrophys. J.* **287**, 571–6.

Miller, R. H. and Prendergast, K. H. (1968). Stellar dynamics in a discrete phase space, *Astrophys. J.,* **151**, 699–701.

Miller, R. H., Prendergast, K. H., and Quirk, W. J. (1970). Numerical experiments on spiral structure, *Astrophys. J.* **161**, 903–16.

Moore, B., Ghigna, S., Governato, R., Lake, G., Quinn, T., and Stadel, J. (1999). Dark matter substructure within galactic halos, *Astrophys. J.* **534**, L19–L22.

Muller, C. A. and Oort, J. H. (1951). Observations of a line in the galactic radio spectrum: the interstellar hydrogen line at 1420 Mc/s and an estimate of galactic rotation, *Nature* **168**, 357.

Navarro, J. F., Frenk, C. S., and White, S. D. M. (1996). The structure of cold dark matter halos, *Astrophys. J.* **463**, 563–75.

Navarro, J. F. and Steinmetz, M. (2000). Dark halo and disk galaxy scaling relations in hierarchical universes, *Astrophys. J.* **538**, 477–88.

Oort, J. H. (1932). The force exerted by the stellar system in the direction perpendicular to the galactic plane and some related problems, *Bull. Astro. Inst. Neth.* **6**, 289–94.

Oort, J. H. (1960). Note on the determination of K_z and on the mass density near the Sun, *Bull. Astro. Inst. Neth.* **494**, 45–63.

Ostriker, J. P. and Peebles, P. J. E. (1973). A numerical study of flattened galaxies: or can cold galaxies survive, *Astrophys. J.* **186**, 467–80.

Ostriker, J. P., Peebles, P. J. E. and Yahil, A. (1974). The size and mass of galaxies and the mass of the Universe, *Astrophys. J.* **193**, L1–L4.

Ostriker, J. P. and Steinhardt, P. J. (1995). The observational case for a low density universe with a non-zero cosmological constant, *Nature* **377**, 600–2.

Paczynski, B. (1987). Giant luminous arcs discovered in two clusters of galaxies, *Nature* **325**, 572.

Peebles, P. J. E. (1965). The black-body radiation content of the Universe and the formation of galaxies, *Astrophys. J.* **142**, 1317–25.

Peebles, P. J. E. (1966). Primordial helium abundance and the primordial fireball II, *Astrophys. J.* **146**, 542–52.

Peebles, P. J. E. (1968). Recombination of the primeval plasma, *Astrophys. J.* **153**, 1–11.

Peebles, P. J. E. (1982). Large scale temperature and mass fluctuations due to scale invariant primeval perturbations, *Astrophys. J.* **263**, L1–L5.

Peebles, P. J. E., Page, L. A., and Partridge, B. (2009). *Finding the Big Bang*, Cambridge University Press (Cambridge).

Penzias, A. A. and Wilson, R. W. (1965). A measurement of excess antenna temperature at 4080 Mc/s, *Astrophys. J.* **142**, 419–421.

Perlmutter, S. (2003). Supernovae, dark energy, and the accelerating universe, *Physics Today* **56**, 53–62.

Perlmutter, S. *et al.* (1997). Measurements of the cosmological parameters omega and lambda from high-redshift supernovae, *Bull. Am. Astron. Soc.* **29**, 1351 (see also arXiv.com, astro-ph/9812473).

Perlmutter, S. *et al.* (1999). Measurements of omega and lambda from 42 high-redshift supernovae, *Astrophys. J.* **517**, 565–86.

Riess, A. *et al.* (2009). Observational evidence from suprenovae for an accelerating Universe and a cosmological constant, *Astron. J.* **116**, 1009–1038.

Roberts, M. S. (1975a). Radio observations of neutral hydrogen in galaxies, *Stars and stellar systems*, Vol. 9 *Galaxies and the Universe*, 309–358.

Roberts, M. S. (1975b). The rotation curves of galaxies, IAU Symp. 69, *The Dynamics of Galaxies*, ed. A. Hayli, Reidel (Dordrecht), pp. 331–339.

Roberts, M. S. and Whitehurst, R. N. (1975). The rotation curve and geometry of M31 at large galactocentric distances, *Astrophys. J.* **201**, 327–46.

Rogstad, D. H. and Shostak, G. S. (1972). Gross properties of five SCD galaxies as determined by 21-centimeter line observations, *Astrophys. J.* **176**, 315–21.

Rood, H. J. (1965). The dynamics of the Coma cluster of galaxies, PhD dissertation, University of Michigan.

Rubin, V. C., Ford, W. K., Thonnard, N. (1980). Rotational properties of 21 SC galaxies with a large range of luminosities and radii, from NGC 4605 (R = 4 kpc) to UGC 2885 (R = 122 kpc), *Astrophys. J.* **238**, 471–87.

Sachs, R. K. and Wolfe, A. M. (1967). Perturbations of a cosmological model and angular variations of the microwave background, *Astrophys. J.* **143**, 73–90.

Sadoulet, B. (2007). Particle dark matter in the Universe: at the brink of discovery?, *Science* **315**, 61–3.

Sancisi, R. (2004). The visible matter – dark matter coupling, IAU Symp. 220, *Dark Matter in Galaxies*, eds. S. D. Ryder, D. J. Pisano, M. A. Walker, and K. C. Freeman, Astron.Soc.Pac. (San Francisco), pp. 233–40.

Sanders, R. H. (1997). A stratified framework for scalar–tensor theories of modified dynamics, *Astrophys. J.* **480**, 492–502.

Sanders, R. H. and Verheijen, M. A. W. (1998). Rotation curves of Ursa Major galaxies in the context of modified Newtonian dynamics, *Astrophys. J.* **503**, 97–108.

Sanders, R. H. and McGaugh, S. S. (2002). Modified Newtonian dynamics as an alternative to dark matter, *Ann. Rev. Astron. Astrophys.* **40**, 263–317.

Schwarzschild M. (1954). Mass distribution and mass-luminosity ratios in galaxies, *Astron. J.* **59**, 273–84.

Schwarzschild, M. and Schwarzschild, B. (1950). A spectroscopic comparison between high and low velocity F dwarfs, *Astrophys. J.* **112**, 248–65.

Seljak, U. and Zaldarriaga, M. (1996). A line-of-sight integration approach to cosmic microwave background anisotropies, *Astrophys. J.* **469**, 437–444.

Sellwood, J. A. (1985). The global stability of our Galaxy, *Mon. Not. Roy. Astron. Soc.* **217**, 127–48.

Shectman, S. A., Landy, S. D., Oemler, A., Tucker, D. L., Lin, H., Kirshner, R. P., and Schechter, P. L. (1996). The Las Companas redshift survey, *Astrophys. J.* **470**, 172–88.

Shostak, G. S. (1973). Aperture synthesis study of neutral hydrogen in NGC 2403 and NGC 4237: II. Discussion, *Astron. Astrophys.* **24**, 411–19.

Shostak, G. S. and Rogstad, D. H. (1973). Aperture synthesis study of neutral hydrogen in NGC 2403 and NGC 4236: I. Observations, *Astron. Astrophys.* **24**, 405–10.

Silk, J. (1967). Fluctuations in the primordial fireball, *Nature* **215**, 1155–6.

Skordis, C., Mota, D. F., Ferreira, P. G., and Boehm, C. (2006). Large scale structure in Bekenstein's theory of relativistic modified Newtonian dynamics, *Phys. Rev. Lett* **96**, 011301.

Smith, S. (1936). The mass of the Virgo cluster, *Astrophys. J.* **83**, 23–30.

Smoot, G. F. *et al.* (1992). Structure in the COBE differential microwave radiometer first-year maps, *Astrophys. J.* **396**, L1–L5.

Soucail, G., Fort, B., Mellier, Y., and Picat, J. P. (1987). A blue ring-like structure in the center of the A 370 cluster of galaxies, *Astron. Astrophys.* **172**, L14–L16.

Spergel, D. N. *et al.* (2007). Three-year Wilkinson microwave anisotropy probe (WMAP) observations: implications for cosmology, *Astrophys. J. Suppl.* **170**, 377–408.

Steinmetz, M. and Navarro, J. F. (1999). The cosmological origin of the Tully–Fisher relation, *Astrophys. J.* **513**, 555–60.

Sunyaev, R. A. and Zeldovich, Ya. B. (1970). Small-scale fluctuations of relic radiation, *Astrophys. Sp. Sci.* **7**, 3–19.

Swaters, R. A. (1999). Dark matter in late-type dwarf galaxies, PhD thesis, University of Groningen.

Szalay, A. S. and Marx, G. (1976). Neutrino rest mass from cosmology, *Astron. Astrophys.* **49**, 437–41.

Tonry, J. L. *et al.* (2003). Cosmological results from high-z supernovae, *Astrophys. J.* **594**, 1–24.

Tremaine, S. and Gunn, J. E. (1979). The dynamical role of light neutral leptons in cosmology, *Phys. Rev. Lett.* **42**, 407–10.

Tully, R. B. and Fisher, J. R. (1977). A new method for determining the distances to galaxies, *Astron. Astrophys.* **54**, 661–73.

Uson, J. M. and Wilkinson, D. T. (1982). Search for small scale anisotropy in the cosmic microwave background, *Phys. Rev. Lett.* **49**, 1463–5.

van Albada, T. S., Bahcall, J. N., Begeman, K., and Sancisi, R. (1985). Distribution of dark matter in the spiral galaxy NGC 3198, *Astrophys. J.* **295**, 305–13.

van Albada, T. S. and Sancisi, R. (1986). Dark matter in spiral galaxies, *Phil. Trans. Roy. Soc.* **320**, 447–64.

van de Hulst, H. C., Raimond, E., and van Woerden, H. (1957). Rotation and density distribution of the Andromeda nebula derived from observations of the 21-cm line, *Bull. Astr. Inst. Neth.* **14**, 1–16.

van der Kruit, P. C. and Searle, L. E. (1981). Surface photometry of edge on spiral galaxies. I. A model for the three-dimensional distribution of light in galactic disks, *Astron. Astrophys.* **95**, 105–15.

Verheijen, M. A. W. (2001). The Ursa Major cluster of galaxies. V. HI rotation curve shapes and the Tully–Fisher relations, *Astrophys. J.* **563**, 694–715.

Vikhlinin, A., Kravtsov, A., Forman, W., Jones, C., Markevitch, M., Murray, S. S. and Van Speybroeck, L. (2006). Chandra sample of nearby relaxed galaxy clusters: mass, gas fraction, and mass–temperature relation, *Astrophys. J.* **640**, 691–709.

Vittorio, N. and Silk, J. (1984). Fine scale anisotropies of the cosmic background radiation in a Universe dominated by cold dark matter, *Astrophys. J.* **285**, L39–L43.

Walsh, D., Carswell, R. F. and Weymann, R. J. (1979). 0957 + 561 A, B - Twin quasistellar objects or gravitational lens, *Nature* **279**, 381–4.

White, S. D. M. (1977). Mass segregation and missing mass in the Coma cluster, *Mon. Not. Roy. Astron. Soc.* **179**, 33–41.

White, S. D. M., Frenk, C. S., and Davis, M. (1983). Clustering in a neutrino dominated Universe, *Astrophys. J.* **274**, L1–L5.

White, S. D. M., Navarro, J. F., Evrard, A. E., and Frenk, C. S. (1993). The baryon content of galaxy clusters: a challenge to cosmological orthodoxy, *Nature* **366**, 429–433.

Zeldovich, Ya. B. (1977). The theory of the large scale structure of the Universe, *Large Scale Structure of the Universe*, IAU Symp. 79, eds. M. Longair and J. Einasto, Reidel (Dordrecht), pp. 409–419.

Zeldovich, Ya. B. and Novikov, I. (1983). Relativistic astrophysics Vol. II: The structure and evolution of the Universe, Univ. of Chicago Press (Chicago).

Zwicky, F. (1933). Der Rotverschiebung von extragalaktischen Neblen, *Act. Helv. Phys.* **6**, 110–127.

Zwicky, F. (1937). On masses of nebulae and clusters of nebulae, *Astrophys. J.* **86**, 217–246.

Index